KENDRICK, RICHARD E
PHYTOCHROME AND PLANT GROWTH
002317926

002317926

QK745.K31

KU-405-410

The Institute of Biology's
Studies in Biology no. 68

UNIVERSITY LIBRARY
LIVERPOOL

Phytochrome and Plant Growth

Richard E. Kendrick

B.Sc., Ph.D.

Lecturer in Plant Biology, Newcastle upon Tyne University

Barry Frankland

B.Sc., Ph.D.

Senior Lecturer in Plant Biology,
Queen Mary College, University of London

Edward Arnold

© R. E. Kendrick and B. Frankland 1976

First published 1976
by Edward Arnold (Publishers) Limited
25 Hill Street, London W1X 8LL.

Board edition ISBN: 0 7131 2560 8
Paper edition ISBN: 0 7131 2561 6

All Rights Reserved. No part of this publication
may be reproduced, stored in a retrieval system,
or transmitted, in any form or by any means, electronic,
mechanical, photocopying, recording or otherwise, without
the prior permission of Edward Arnold (Publishers) Limited.

Printed in Great Britain by
The Camelot Press Ltd, Southampton

General Preface to the Series

It is no longer possible for one textbook to cover the whole field of Biology and to remain sufficiently up to date. At the same time teachers and students at school, college or university need to keep abreast of recent trends and know where significant developments are taking place.

To meet the need for this progressive approach the Institute of Biology has for some years sponsored this series of booklets dealing with subjects specially selected by a panel of editors. The enthusiastic acceptance of the series by teachers and students at school, college and university shows the usefulness of the books in providing a clear and up to date coverage of topics, particularly in areas of research and changing views.

Among features of the series are the attention given to methods, the inclusion of a selected list of books for further reading and, wherever possible, suggestions for practical work.

Readers' comments will be welcomed by the authors or the Education Officer of the Institute.

The Institute of Biology, 41 Queens Gate,
London, SW7 5HU

1976

Preface

Green plants, and ultimately all forms of life, depend upon the utilization of sunlight as an energy source in the process called photosynthesis. However, the intensity, quality and duration of light can regulate the growth and development of a plant independently of photosynthesis. These responses of a plant to light are collectively referred to as *photomorphogenesis*.

The discovery and isolation of phytochrome, the red/far-red reversible photoreceptor involved in photomorphogenesis, is one of the most exciting chapters in modern biology. It provides an excellent illustration of the nature of the scientific process and the way in which scientific ideas develop.

The information in this book is presented in a compact and concise form which should make it suitable as advanced reading for the sixth form student. It is hoped that it is also sufficiently complete and up to date to provide background reading for the undergraduate student taking a course in, say, plant physiology.

From the point of view of the biology student the study of plant photomorphogenesis is not only important as an aspect of plant physiology, plant biochemistry or plant ecology but can be used to illustrate general principles of photobiology and developmental biology.

Newcastle upon Tyne and   R.E.K.
London, 1976   B.F.

353505

Contents

1 Introduction   1
1.1 Light and the regulation of plant growth   1.2 Red/far-red
reversibility   1.3 Absorption spectra   1.4 Action spectra

2 Phytochrome Detection and Isolation   12
2.1 Historical aspects   2.2 Spectrophotometry of optically
dense samples   2.3 The spectrophotometric assay   2.4 Distribution and localization   2.5 Extraction and purification

3 The Properties of Phytochrome   24
3.1 Photoconversion   3.2 Dark transformations   3.3 Biosynthesis   3.4 Phylogenetic aspects

4 Phytochrome Controlled Responses   37
4.1 Photoresponses   4.2 The classical red/far-red reversible
reaction   4.3 Rate of response   4.4 Relationship between
response and incident light energy   4.5 Is Pfr the physiologically
active form of phytochrome?   4.6 Relationship between
response and Pfr/P ratio   4.7 Transmissible phytochrome
effects   4.8 Escape from reversibility: time course of Pfr action
4.9 Apparent incomplete reversal by far-red   4.10 Requirement for repeated irradiation with red light   4.11 Far-red
inhibition of seed germination   4.12 Phytochrome action in
light-grown plants   4.13 Phytochrome and biological time
measurement   4.14 The high irradiance reaction   4.15 Phytochrome and plants in their natural environment

5 Mode of Action of Phytochrome   53
5.1 Analysis of Pfr action   5.2 Phytochrome and gene expression
5.3 Phytochrome and enzyme synthesis   5.4 Phytochrome
and permeability   5.5 Intracellular localization of phytochrome
5.6 Phytochrome in membranes   5.7 Phytochrome and the
future

6 Some Practical Exercises   63
6.1 Construction of simple light sources   6.2 Phytochrome
control of seed germination   6.3 Phytochrome control of
seedling growth   6.4 Phytochrome control of carbohydrate
metabolism

Further Reading   67

1 Introduction

1.1 Light and the regulation of plant growth

Light is the energy source on which plants and ultimately all living things depend. However, in addition to its utilization in the process of photosynthesis, light can play an important regulatory role in plant growth. Higher animals perceive their surroundings and respond to them through the interaction of eyes, brain and limbs, the visual pigment of the eye being the primary light detector. Although plants, unlike most animals, are sedentary organisms they can be orientated by growth movements which are directionally related to the light source. Such responses maximize the efficiency of light interception by the leaves. Darwin, in his book *The Power of Movement in Plants* published in 1880, first fully documented the ability of plants to grow towards the light. The receptor pigment involved in these *phototropic* movements has yet to be identified.

In addition to directing such tropic movements light controls developmental processes such as seed germination, seedling development and flowering. A number of questions can be asked about these events in the life of a plant. How is it that a plant flowers at a particular time of year? How is it that a seedling ceases rapid extension growth on penetrating the soil surface and develops leaves for photosynthesis before its limited food reserves have been exhausted? How is it that some seeds do not germinate while buried in the soil but do germinate when exposed at the soil surface by cultivation? The questions can be rephrased to elicit different kinds of answers. What is the environmental stimulus involved? How is the stimulus perceived? What are the processes involved in producing the appropriate response?

In the case of seed germination and seedling development the environmental stimulus is simply the presence of light. Plants have evolved a mechanism for detecting radiant energy, the ability to do so clearly conferring a selective advantage. The regulation of plant growth and development in relation to the light environment is of obvious importance to a sedentary organism dependent on light as its energy source. Responses are not only to the presence or absence of light but also to variation in the intensity of light. For instance, plants growing in a shaded situation tend to be taller and have a greater leaf area to plant weight ratio than those growing in full sunlight. Responses may also be observed at the sub-cellular level. The disc-shaped chloroplasts of many plants become orientated so that their maximum cross-section is towards the light at low light intensities allowing maximum light absorption. At very high light intensities they become orientated so that their minimum cross-section is towards the light thus protecting the photosynthetic pigments from damage.

Light duration is also an important regulatory factor. For instance, the length of the day is the environmental factor which provides the most accurate indication of the time of year and plants may respond to a change in the length of the day with a dramatic change in their pattern of development. This phenomenon is termed *photoperiodism*. The best known example is the induction of flowering. Other examples are the onset of bud formation and leaf fall in deciduous trees in response to the short days of autumn.

These various *formative* effects of light are collectively referred to as *photomorphogenesis*. Although they have been known for a long time, it was not until 1935 that any advance was made towards the discovery of the pigment or photoreceptor involved. Flint and McAlister working with lettuce seeds found that the wavelengths of light most effective in promoting germination were in the red region of the spectrum. They also showed that germination was depressed, below that in the dark, by light in the far-red region of the spectrum. In 1945 a large spectrograph was constructed at the Plant Industry Station of the United States Department of Agriculture, Beltsville, Maryland. This enabled Borthwick and Hendricks, and their co-workers, to determine the relative effectiveness of different wavelengths of light in bringing about particular responses, i.e. action spectra (see 1.4).

In 1922 Garner and Allard discovered that the length of the day was the critical factor in the control of flowering. They distinguished between short day plants (SDP) which flower when days become shorter than a critical length in late summer and long day plants (LDP) which flower as days grow longer than a critical length in early summer. They further showed that it was the length of the dark period rather than the length of the light period which was measured by the plant. A SDP such as soybean or cocklebur could be prevented from flowering under short days by exposing it to light for a few minutes in the middle of the night (Fig. 1–1). Clearly the amount of light energy involved here is too small for the effects to be accounted for in terms of difference in photosynthesis. The action spectrum for this response pointed to a receptor pigment with an absorption spectrum (see 1.3) significantly different from that of chlorophyll, although having a peak in the red region of the spectrum. A similar experiment was carried out with the long day plants barley and henbane under short days but in this case a light break given in the middle of the night induces flowering (Fig. 1–1). Action spectra for this response showed a close similarity to those for the inhibition of flowering in SDP.

Seedlings which are grown in the dark show characteristic symptoms such as elongated stems and poor leaf development. These symptoms are collectively referred to as *etiolation* and were studied as early as 1754 by Bonnet. Short irradiations with light are sufficient for de-etiolation.

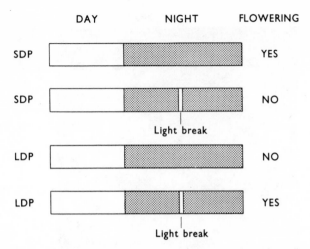

Fig. 1–1 The effect of a light break in the middle of the night on flowering of short day plants (SDP) and long day plants (LDP) when they are grown under short days.

Action spectra of such responses (e.g. the promotion of leaf development in pea seedlings) were shown to be very similar to those found in the case of flowering, indicating a common photoreceptor. Responses of albino plants, i.e. plants lacking chlorophyll, proved to be the same as those of normal green plants, confirming the suggestion that the photosynthetic system was not involved.

1.2 Red/far-red reversibility

Detailed action spectra were determined for the promotion of lettuce seed germination and shown to be similar to those determined in flowering and de-etiolation experiments. The wavelength of maximum effectiveness was 660 nm (red light). An action spectrum of the inhibition of the 25% of lettuce seeds that germinate in darkness showed a peak of effectiveness at wavelength 730 nm (far-red light). A most significant observation was that far-red light not only inhibited germination but could reverse the promoting effect of a previous irradiation with red light (Fig. 1–2). The reversal by far-red light of the effect of red light can be repeated many times, the germination response depending on the final wavelength of light given. It was concluded that the promotion and inhibition of germination by red and far-red light were mediated by a single pigment.

Since the action spectrum for promotion of germination coincided with those for flowering and seedling photomorphogensis attempts were made to see if far-red light could also reverse the effects of red light in

4

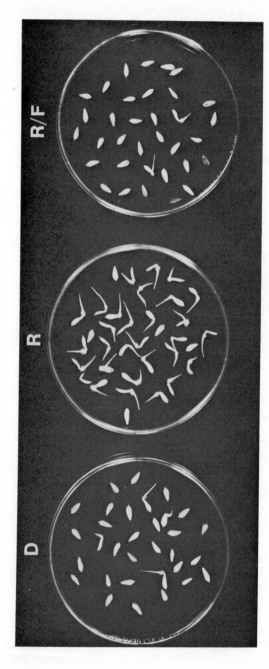

Fig. 1-2 Red/far-red reversibility in light stimulated germination of lettuce seeds. D, dark; R, received 3 minutes red light; R/F, received 3 minutes red light followed by 3 minutes far-red light.

these cases. Eventually red/far-red reversibility was demonstrated for flowering and many de-etiolation phenomena. Typical action spectra for the opposing effects of red and far-red light are shown in Fig. 1–3. In 1952 Borthwick and Hendricks proposed that a red/far-red reversible pigment

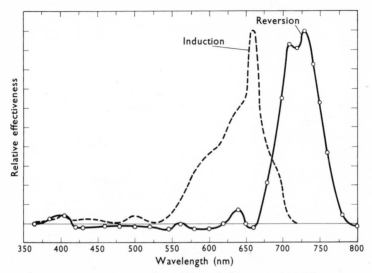

Fig. 1–3   Typical action spectra for red/far-red photoresponses. The spectra are for the induction and reversion of plumular hook opening of etiolated bean seedlings. They are very similar to action spectra of other red/far-red reversible responses, such as promotion of lettuce seed germination. (Redrawn from WITHROW, R. B., KLEIN, W. H. and ELSTAD, V. (1957). *Plant Physiol.* **32**, 453.)

existed in the plant which had an inactive red-absorbing form (called Pr) and an active far-red absorbing form (called Pfr).

$$Pr \xrightleftharpoons[\text{far-red}]{\text{red}} Pfr$$

Action spectra were taken as an indication of the absorption spectrum of the pigment, suggesting that it was blue-green in colour.

1.3   Absorption spectra

Light is a form of electromagnetic radiation. In common with other electromagnetic radiations it travels at a velocity $c = 3 \times 10^8 \text{ms}^{-1}$. The electromagnetic spectrum, ranging from very short wavelength (λ) cosmic rays ($\lambda < 10^{-12}$ m) to long wavelength radio waves ($\lambda > 10^{-1}$ m) is shown in Fig. 1–4. Visible light is that portion of the spectrum that can be detected by the human eye, i.e. of wavelength between 380 and 750 nm (1

6

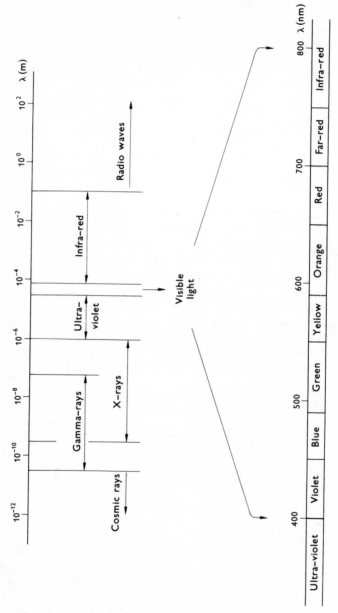

Fig. 1-4   The electromagnetic spectrum.

nm $= 10^{-9}$ m). Visible light is bounded by the near ultra-violet on the short wavelength side and the near infra-red on the long wavelength side. Most organic compounds in a plant absorb ultra-violet light but a few, called pigments, also absorb in the visible region of the spectrum by virtue of their extended π-electron systems.

The Quantum theory of radiation transfer as proposed by Planck states that transfer of radiation takes place in discrete packets of energy which he called *quanta*. Mathematically this was written:

$$E = h\nu$$

Where E is the energy of a single quantum, h is Planck's constant and ν is the frequency. Later Einstein extended this theory to light and called the energy of a single quantum of light a *photon*. Since frequency is inversely related to wavelength it follows

$$E = h\frac{c}{\lambda}$$

Photons of short wavelength are more energetic than photons of longer wavelength, i.e. a photon of blue light contains more energy than a photon of red light.

For a single molecule to undergo a photoreaction it must absorb one quantum of energy $(h\nu)$. Therefore one mole of a compound would absorb $N(N=6.024 \times 10^{23}$, the Avogadro number) photons of energy $Nh\nu$. The total energy of the photons absorbed by a mole of a compound is called an Einstein (E). In biological systems the dose of a wavelength of light is expressed in Einsteins, being a measure of the number of quanta involved in a photoreaction. The energy of an Einstein of red light is less than an Einstein of blue light, although both have the same number of quanta.

Light is usually measured by an instrument, such as a radiometer, whose response is proportional to the radiant energy, rather than the number of photons, falling on the surface of the detector. Therefore, biologists very frequently express light doses in terms of energy per unit area, e.g. J (Joules) m^{-2}. Light intensity, or what is more properly called *irradiance*, is expressed as energy per unit area per unit time, i.e. J m^{-2} s^{-1}, or power per unit area i.e. W (Watts) m^{-2}. The illumination of a surface by white light is often measured by a photometer in units of *lux* (lumens m^{-2}). Such units cannot be converted directly into irradiance units and can only be used for comparing light sources of a similar spectral quality. 1 lux at 555 nm is equivalent to 1.61 mW m^{-2}. The Quantum equivalent of irradiance is called quantum flux density (e.g. nanoEinsteins (nE) m^{-2} s^{-1}).

When a molecule absorbs a quantum of light its energy is increased. This is represented by the excitation of an electron to a higher energy level. Two electrons cannot absorb one photon nor can one photon excite two electrons. When an electron absorbs a photon it is said to be in an excited state as opposed to the ground state. For individual atoms the absorption of a single quantum of energy can be followed by a sharp line in its absorption spectrum at the wavelength λ given by $\Delta E = hc/\lambda$. In a molecule consisting of different atoms the transitions from ground to the excited state can take place by absorption of light quanta of varying amounts of energy, the sharp line in the absorption spectrum being replaced by a broader absorption band. An excited state can revert to the stable ground state with the emission of a quantum of light. Whereas with an atom the absorption and emission takes place at the same wavelength, in a molecule the peak of the emission (e.g. fluorescence) spectrum is at a longer wavelength (i.e. lower energy) than that of the absorption spectrum. The absorption spectrum is characteristic of a molecule and it gives an indication of the nature of its chemical structure. Coloured organic compounds, i.e. those that absorb visible light, have highly conjugated systems of π electrons. An example is the porphyrin ring of chlorophylls.

The absorption spectrum (Fig. 1–5) can be defined as a graph of absorbance (A) of a pigment plotted against wavelength (λ). The

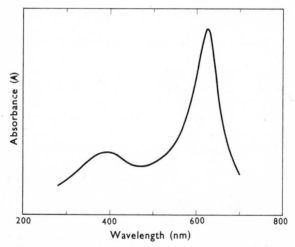

Fig. 1–5   The absorption spectrum of a hypothetical pigment. Absorbance plotted against wavelength.

absorbance of a pigment is measured by a spectrophotometer. This instrument provides monochromatic light of various wavelengths at which the absorbance can be measured. This has to be carried out under certain standard conditions (Fig. 1–6). Pigment solution is placed in a cuvette through which the light is passed. Correction has to be made for light reflected at the surface of the cuvette and light absorbed by the

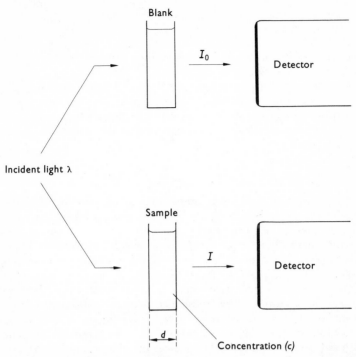

Fig. 1–6 The principle of absorbance measurement using a spectrophotometer. Arrows indicate the direction of the monochromatic light beam. I_0, is the light intensity transmitted by the reference or blank cuvette; I, is the light intensity transmitted by the sample cuvette; d, is path length and c, the pigment concentration.

solvent. This is done by comparing with an identical cuvette holding the same solvent but lacking the pigment under investigation (called the blank or reference cuvette). The intensity of the light emerging from the cuvette is measured by means of a photosensitive device such as a photomultiplier tube.

Beer's law states that the absorbance (A) is proportional to the concentration (c) of the pigment:

$$A = \log_{10}\frac{I_0}{I_0} \propto c \ \text{(if the pathlength } (d) \text{ is constant)}$$

where I_0 is the intensity of light from the blank cuvette and I is the intensity of light from the sample cuvette.

Also:

$$A \propto d \ \text{(if the concentration } (c) \text{ is constant)}$$

Therefore:

$$A = \epsilon\,cd$$

where ϵ is the combined proportionality constant and is called the extinction coefficient.

Therefore $\epsilon = A$ for 1M solution of 1 cm path length and is called the molar extinction coefficient. ϵ plotted against wavelength gives the absorption spectrum of the pigment. The absorption spectrum of a pigment is characteristic of the pigment and indicative of a particular chemical structure.

1.4 Action spectra

An action spectrum is a graph of the effectiveness of different wavelengths of light in bringing about a particular biological response. As seen in section 1.3 a response initiated by a photoreaction requires the absorption of discrete quanta. The first step in determining an action spectrum is to determine the number of quanta of different wavelengths required to bring about the response. This involves plotting dose response curves for each wavelength. It is important to plot the response against the quantum dose (Einsteins), because as pointed out earlier quanta of different wavelengths contain different amounts of energy. A standard response is chosen (X in Fig. 1–7) and the number of quanta of each wavelength required to bring about this response are measured. The lower the number of quanta of a particular wavelength required the more effective that wavelength is in bringing about the reaction. A plot of $1/N$,

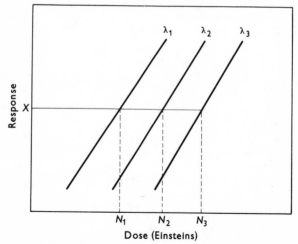

Fig. 1–7 Dose response curves enable the dose (N) of different wavelengths of light ($\lambda_1, \lambda_2, \lambda_3$) required to bring about a standard response (X) to be determined.

where N = dose to bring about the standard response, against λ (wavelength) is called the action spectrum (Fig. 1–8). It should be the same shape as the absorption spectrum of the receptor pigment (Fig. 1–5), providing screening pigments are not present.

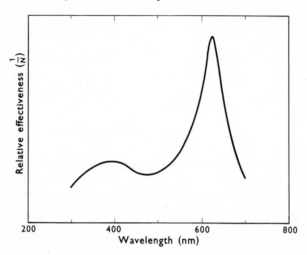

Fig. 1–8 The action spectrum of a hypothetical photoresponse. Relative effectiveness $(1/N)$ in bringing about a standard response plotted against wavelength. In this case the receptor pigment is the same as that whose absorption spectrum is shown in Fig. 1–5.

An action spectrum provides a powerful tool in determining the nature of a photoreceptor involved in a biological response. This is especially true when the pigment is very dilute and not easily seen. Since an action spectrum has the same form as the absorption spectrum of the receptor pigment it gives an indication of the chemical structure of a photoreceptor even before the latter can be physically detected.

2 Phytochrome Detection and Isolation

2.1 Historical aspects

On the basis of action spectra for red/far-red reversible responses in plants it was predicted that there was a photoreversible pigment existing in two forms, Pr (red absorbing) and Pfr (far-red absorbing), which were respectively blue and green in colour. Action spectra suggested that the

$$Pr \xrightleftharpoons[\text{far-red}]{\text{red}} Pfr$$

absorption spectrum of the red-absorbing form of the pigment was similar to that of the blue-green algal pigment c-phycocyanin. This is known to consist of an open-chain tetrapyrrole chromophore attached to a protein (Fig. 2–1). In 1952 the presence of such a pigment in a plant could only be inferred from the red/far-red reversible nature of a physiological response. At this time attempts were made to detect

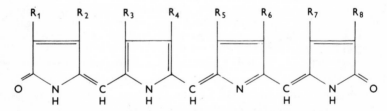

Fig. 2–1   General structure of an open-chain tetrapyrrole.

phytochrome by physical means. Seedlings grown in the dark and albino maize plants were first examined, since these plants lack chlorophyll which would obviously mask any other green pigment. However, such seedlings are yellow rather than blue or green in colour despite the fact that they exhibit red/far-red reversible responses. This indicated that the pigment, if present, was in a very low concentration.

Alternating exposure to red and far-red light should induce changes in absorbance in a sample containing the pigment, and a properly designed and sufficiently sensitive spectrophotometer should be able to detect such changes. A sample consisting of plant tissue is very different from a clear solution (see 1.3). The samples are optically dense because they exhibit light scatter brought about by variation in refractive index throughout the tissue (see 2.2). An instrument was already available at Beltsville for work on optically dense material, and in 1959 it was used by Butler and co-workers to look for differences in absorption spectra between red-irradiated and far-red irradiated dark-grown maize seedling tissue. Differences were immediately found (Fig. 2–2) with the maximum fall in absorbance being at 660 nm and the maximum rise in absorbance being at 730 nm in going from far-red irradiated to red irradiated tissue. The photomorphogenetic pigment was now physically detected, and it was at this time that the name *phytochrome* (meaning 'plant pigment') was first proposed by Butler. Within hours of it being detected in tissue by spectrophotometry an aqueous extract was prepared and shown to contain phytochrome. The photoreversible properties were lost on boiling, suggesting that the phytochrome chromophore was attached to a

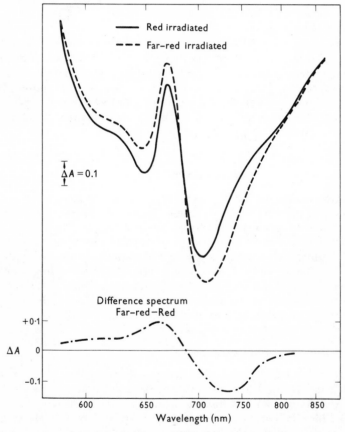

Fig. 2–2 First reported 'difference spectrum' for phytochrome. The upper curves are the actual absorbance after saturating exposures to red and far-red light and the difference is shown below. (Redrawn from BUTLER, W. L., NORRIS, K. H., SIEGELMAN, H. W. and HENDRICKS, S. B. (1959). *Proc. Nat. Acad. Sci. U.S.A.* 45, 1703.)

protein. In 1964 Siegelman and Firer used this knowledge to develop a method of purification of phytochrome from etiolated seedlings (see 2.5).

2.2   Spectrophotometry of optically dense samples

Unlike clear solutions optically dense samples do not obey Beer's Law (see 1.3). When a sample is placed in a spectrophotometer most of the light falling on the cuvette is lost because of scatter. That light which emerges from the rear of the cuvette does so in all directions. The sensitivity of a spectrophotometer depends on the signal-to-noise ratio, S/N

$$\frac{S}{N} \propto I$$

where I=light flux falling on the detector (photomultiplier). It is therefore important with an optically dense sample to have measuring beams of high intensity and to collect as large a solid angle of the emergent light as possible by placing the detector very close to the sample. When measuring the products of a reversible photoreaction the measuring beams cannot be of too high an intensity or they will bring about pigment photoconversion. For a light scattering sample:

$$\text{apparent absorbance}\,(A) = \text{scatter} + \text{real absorbance}$$

$$\text{real absorbance} = \beta\,\epsilon\,c\,d$$

where β=scattering factor; ϵ=extinction coefficient; c=average concentration of pigment; d=path length (sample thickness). The scattering essentially increases the path length by producing a tortuous path along which a quantum of light has greater probability of being absorbed. Other things being equal the presence of scatter gives an amplification of the absorbance by increasing path length. This factor is important and makes detection of low pigment concentrations possible. An assay system which measures in an optically dense system, must in some way compensate for that fraction of the light loss due to scatter.

2.3   The spectrophotometric assay

The most widely used assay for phytochrome has been the dual wavelength method. This involves measuring the apparent absorbance at two wavelengths and subtracting them to give an absorbance difference (ΔA). This difference is a measure of the real absorbance difference at the two wavelengths since the scatter components of the apparent absorbances at the two wavelengths are equal and therefore cancel out. In the case of phytochrome the wavelengths used depend on the material concerned. Initially the beams are set at the peak absorbances of Pr and

Pfr, 660 and 730 nm respectively. The ΔA is then measured after sequential irradiations with red and far-red light to bring about phytochrome conversion (Fig. 2–3). Obviously, using these wavelengths changes in ΔA after red light will arise from both a loss of Pr and a gain of

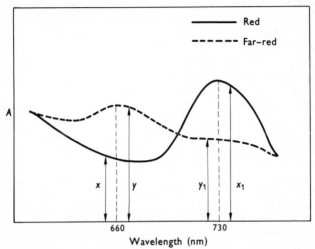

Fig. 2–3   Diagram of the absorbance changes measured in the phytochrome assay.

Pfr. Therefore the magnitude of the ΔA readings are aproximately twice that corresponding to the amount of phytochrome present.

$$\Delta A\text{fr} = y{-}y_1 \text{ (after far-red light)}$$
$$\Delta A\text{r} = x{-}x_1 \text{ (after red light)}$$
$$\text{Therefore total phytochrome} = \Delta(\Delta A) = (y{-}y_1) - (x{-}x_1) = \Delta A\text{fr}{-}\Delta A\text{r}$$
$$\Delta(\Delta A) = 2\,\beta ecd.$$

Pr and Pfr can be measured separately if the wavelengths 660 and 730 nm are used independently with a reference wavelength of 800 nm at which phytochrome has little, if any, absorbance.

The *in vivo* assay cannot be used with tissues that contain more than a small amount of chlorophyll as the measuring light beams are absorbed by chlorophyll as well as by phytochrome. This precludes measurement in green leaves, although phytochrome is detectable in achlorophyllous tissues from plants grown in the light such as white flower petals and white regions of variegated leaves. This means that most of our knowledge of phytochrome *in vivo* comes from studies on etiolated tissues which are rich in phytochrome and low in chlorophyll. However, tissues grown in the dark often contain a little of the precursor of chlorophyll, protochlorophyllide. When irradiated with red light this undergoes an irreversible photoreaction to form chlorophyllide, which absorbs at

longer wavelengths. At room temperature there is a gradual absorbance
shift from longer to shorter wavelength as chlorophyll is formed from
chlorophyllide. The light reaction is very rapid and the dark reactions
take about 20 minutes to reach completion at room temperature. At the
temperature used for the phytochrome assay ($0°C$) these dark shifts do not
occur. Exposure of tissue grown in the dark to red light results in an
absorbance increase at 660 nm due to protochlorophyllide conversion
and an absorbance decrease at 660 nm due to phytochrome conversion.
Often this increase and decrease are about equal. This problem can be
partially overcome by giving irradiation with red light prior to the
red/far-red/red sequence used to determine the amount of phytochrome
in a sample. Chlorophyll artefacts are most noticeable using measuring
beam wavelengths of 660 and 730 nm, but can be minimized by using 730
versus 800 nm. In the latter case, of course, the values obtained are about
half of those obtained with the 660/730 measuring system (Table 2.1).

Table 2.1   Comparison of phytochrome measurements in dark-grown seedlings
of *Amaranthus* using measuring beam wavelengths of 730 versus 660 nm and 730
versus 800 nm. The initial D values were set at an arbitrary value of $20 \times 10^{-3}\Delta A$
in both cases. F, 2 minutes far-red light; R, 2 minutes red light.

Irradiation sequence	$10^3\,\Delta A$ 660–730 nm	$10^3\,\Delta A$ 730–800 nm
D	20	20
F	21	20
R	17	40
F	48	24
R	23	41
F	52	25
R	26	42
F	55	26
Mean $\Delta(\Delta A)$	30	16

Figure 2–4 shows the principle of operation of a dual wavelength
spectrophotometer. The exact construction varies considerably, but a
simple version is shown in Fig. 2–5. This spectrophotometer uses
interference filters to obtain the measuring beam wavelengths λ_1 and λ_2
which are sequentially irradiated on to the sample by means of a rotating
sectioned mirror. Provision is made for the introduction of a prism and
photomultiplier shutter so that the sample can be irradiated with actinic
light to photoconvert the phytochrome.

Although the dual-wavelength method is the most sensitive it only
enables individual wavelengths to be examined. To obtain a complete
'difference spectrum' for the reaction Pr $\xrightarrow{h\nu}$ Pfr it is necessary to
make measurements at many wavelengths. A second method is available

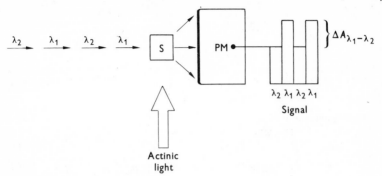

Fig. 2-4 Schematic diagram of a dual wavelength spectrophotometer. The measuring beams of wavelength λ_1 and λ_2 fall successively on to the sample cuvette (S). The photomultiplier (PM) measures the light of each wavelength emerging from the sample and the corresponding signals are compared to give a direct absorbance difference $\Delta A_{\lambda_1-\lambda_2}$. Also provision is made for photoconverting the pigment in the sample cuvette with actinic light.

which enables the whole spectrum to be scanned, but this is less sensitive and requires two samples. This is the double beam technique. As shown in Fig. 2-6 this method involves measuring the difference in absorbance between the two samples before and after photoconversion of phytochrome in one of them from Pr to Pfr. The instrument gives directly the difference spectrum, i.e. the difference between a reference cuvette containing Pr and a sample cuvette containing Pfr. Since in other respects the two samples are equal the difference should reflect only photoconversion of phytochrome.

2.4 Distribution and localization

Phytochrome has now been detected in achlorophyllous tissue of many species. Etiolated seedlings grown in the dark have been used extensively. Detailed studies of different parts of seedlings suggest that the highest phytochrome content (about 10^{-6} M) is in the regions of most active growth such as the epicotyl hooks of peas and the coleoptilar node of oats (Fig. 2-7). It has been detected directly in roots, hypocotyls, cotyledons, coleoptiles, stems, petioles, leaf blades, vegetative buds, floral receptacles, inflorescences and developing fruits and seeds. It has been detected in tissues of both primary and secondary growth. Besides these direct observations indirect evidence from physiological experiments indicates the presence of phytochrome in many other plant materials. Phytochrome has been detected in angiosperms, gymnosperms, bryophytes and algae. Differences between phytochrome in higher and lower plants are discussed in section 3.4.

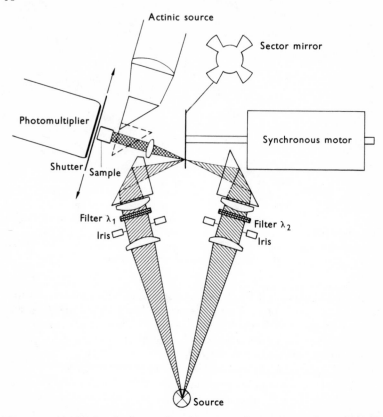

Fig. 2–5 A simple dual wavelength spectrophotometer. The measuring beams of wavelength λ_1 and λ_2 are obtained by means of interference filters from the same source and they fall successively on the sample by means of a rotating sectored mirror. Iris diaphragms enable the beams to be balanced. Insertion of a prism and closing a protective shutter in front of the photomultiplier enable the sample to be irradiated with actinic light. (Redrawn from SPRUIT, C. J. P. (1970). *Meded. Landbouwhogeschool Wageningen*, **70**, 1.)

2.5 Extraction and purification

The most satisfactory plant material for the extraction of phytochrome has proved to be dark-grown seedlings of monocotyledons such as oats and rye. Phytochrome can be extracted from tissues of green plants but only with considerable difficulty.

Phytochrome is readily extracted from oats by grinding dark-grown coleoptile tissue with buffer at a pH higher than 7.3. Below this pH the phytochrome is found in the pellet following centrifugation to remove cell debris. Since phytochrome is only present at low concentration large

19

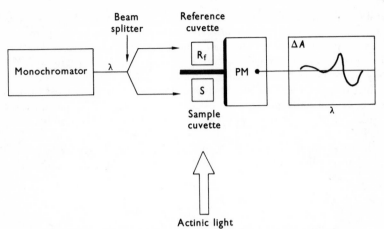

Fig. 2–6 Schematic diagram of a double beam spectrophotometer. A monochromatic light beam of wavelength λ is split so that it falls successively on to the reference cuvette (R_f) and sample cuvette (S). The light emerging from the two cuvettes is measured by the photomultiplier (PM) and the two signals are adjusted electronically to be equal throughout the spectrum (base line). Only the sample cuvette is then exposed to actinic light and the spectrum scanned to give directly the difference spectrum of the photoreaction.

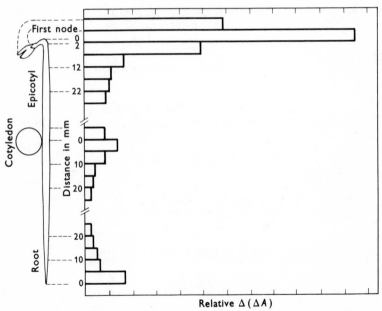

Fig. 2–7 Distribution of phytochrome in an etiolated pea seedling. (Redrawn from FURUYA, M. and HILLMAN, W. S. (1964). *Planta*, **63**, 31.)

volumes of tissue and extraction buffer are required. Since it is a protein and therefore labile, especially in crude extracts, low temperatures (0–2°C) are used and a reducing agent such as mercaptoethanol added. The extraction is carried out under a dim green safe light with the phytochrome in the more stable Pr form.

Figure 2–8 shows a typical scheme for the purification of phytochrome. It involves initially grinding the tissue and reducing the large volumes of crude extract to manageable proportions. This is followed by many purification steps including adsorption chromatography, ammonium sulphate precipitation, gel filtration and ion exchange chromatography to separate phytochrome from the large number of proteins of similar

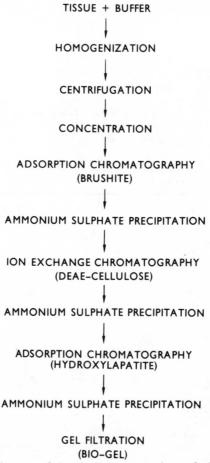

TISSUE + BUFFER

HOMOGENIZATION

CENTRIFUGATION

CONCENTRATION

ADSORPTION CHROMATOGRAPHY
(BRUSHITE)

AMMONIUM SULPHATE PRECIPITATION

ION EXCHANGE CHROMATOGRAPHY
(DEAE–CELLULOSE)

AMMONIUM SULPHATE PRECIPITATION

ADSORPTION CHROMATOGRAPHY
(HYDROXYLAPATITE)

AMMONIUM SULPHATE PRECIPITATION

GEL FILTRATION
(BIO–GEL)

Fig. 2–8   Flow diagram of the extraction procedure of phytochrome from etiolated oat seedlings.

charge and size. The absorption spectrum of phytochrome in its two forms is shown in Fig. 2–9. The colour was observed to change from blue-green (Pr) after irradiation with far-red light to green (Pfr) after irradiation with red light. After action spectra were determined for photoconversion it became clear that red light can also photoconvert Pfr to Pr to some extent. This means that a saturating irradiation with red light produces a mixture of 80% Pfr and 20% Pr. Far-red light produces approximately 3% Pfr and 97% Pr.

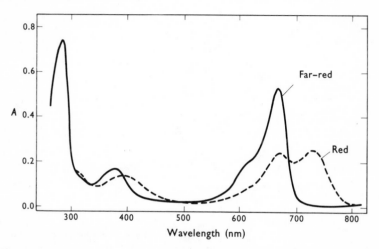

Fig. 2–9 The absorption spectrum of a solution of phytochrome after a saturating exposure to red and far-red light. (Redrawn from RICE, H. V., BRIGGS, W. R. and JACKSON-WHITE, C. J. (1973). *Plant Physiol.* **51**, 917.

The action spectra of physiological responses led Hendricks and his co-workers to speculate that the tetrapyrrole chromophore of phytochrome was similar to that of the photosynthetic pigment c-phycocyanin of blue-green algae. The absorption spectrum of purified phytochrome confirmed this view and it was proposed that the chromophore group which gives phytochrome its colour was of the type in Fig. 2–1. The exact nature of the chromophore has not yet been determined, mainly because of the difficulty in obtaining the large quantities required for chemical analysis. However, most of the groups R_1 to R_8 are now known and suggestions of the nature of the isomerization that occurs on irradiation with red and far-red light have been put forward (Fig. 2–10). The absorption spectrum of the phytochrome chromophore is shown in Fig. 2–11 and is compared with the chromophore of c-phycocyanin.

Over the years the molecular weight of oat phytochrome became accepted as 60 000 daltons in most laboratories. However, Correll using rye as a source obtained a phytochrome preparation of high purity with a

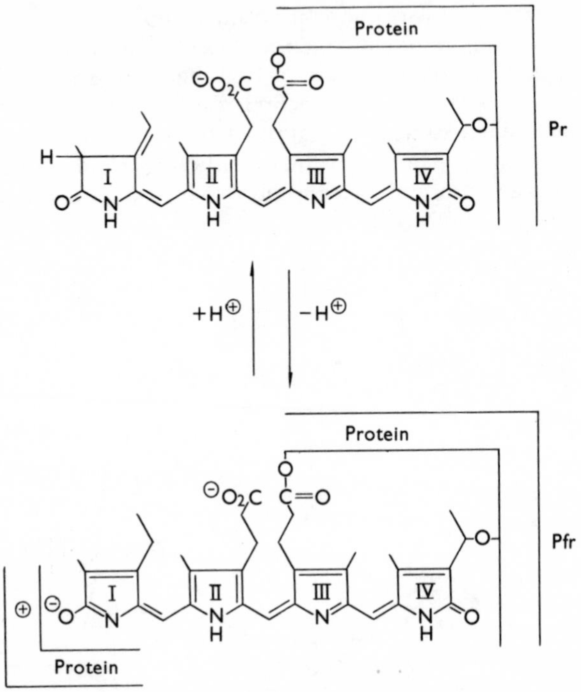

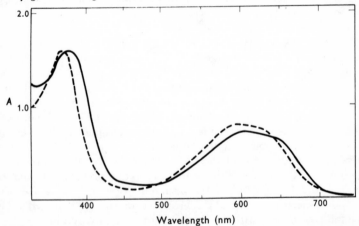

Fig. 2–10 Proposed structure of the phytochrome chromophore and its linkage to protein. (Redrawn from RUDIGER, W. and CORRELL, D. L. (1969). *Liebigs Ann. Chem.* **723**, 208. Verlag Chemie, GmbH., Weinheim/Germany.)

Fig. 2–11 Comparison of the absorption spectrum of the isolated chromophores of phytochrome (———) and c-phycocyanin (- - - -). (Redrawn from KROES, H. H. (1970). *Meded. Landbouwhogeschool Wageningen*, **70**, 1.)

higher molecular weight which on storage gave a product with lower molecular weight. Briggs and co-workers in 1972 obtained a partially pure phytochrome preparation from oats with components of 120 000 and 60 000 molecular weight. After storage for a day in the cold (0–4°C) only the smaller component was found. It was demonstrated that a protease was present in oats which attacked the phytochrome during the early stages of the extraction procedure. This protease degrades the larger molecular weight form of phytochrome to a stable 60 000 molecular weight product. Rye contains far less of this endogenous protease and it is therefore simpler to obtain large phytochrome from this source. The nature of the protease was investigated and the extraction procedure was modified to minimize its action.

An amino acid analysis of 120 000 molecular weight phytochrome is shown in Table 2.2. Recent immunochemical studies which involve making antisera of large and small phytochrome suggest that when large phytochrome is degraded not only is the 60 000 molecular weight red/far-red reversible form produced, but also a protein that is not photoreversible. It is therefore possible that there is only one chromophore per 120 000 molecular weight form.

Table 2.2 Amino acid composition of 120 000 molecular weight rye phytochrome. (From RICE, H. V. and BRIGGS, W. R. (1973), *Plant Physiol.*, **52**, 927.)

Amino acid	Number of moles
Lysine	58
Histidine	28
Arginine	47
Aspartic acid	104
Threonine	46
Serine	75
Glutamic acid	128
Proline	88
Glycine	77
Alanine	110
Cystine	52
Valine	89
Methionine	32
Isoleucine	54
Leucine	111
Tyrosine	23
Phenylalanine	43
Total =	1165

3 The Properties of Phytochrome

3.1 Photoconversion

The photoconversion of Pr → Pfr and of Pfr → Pr are both first order reactions (Fig. 3–1). This means that, at constant irradiance (I), the rate of conversion of Pr to Pfr at a given time (t) is proportional to the amount of Pr remaining at that time.

$$\frac{\mathrm{dPr}}{\mathrm{d}t} = -x.I.\mathrm{Pr}$$

It also follows that there is a linear relationship between the logarithm of Pr remaining and the light dose ($I.t$) given. The gradient of the line is the action constant (x). The action constant is the product of the extinction coefficient and the quantum efficiency (see 1.3). Being a photochemical

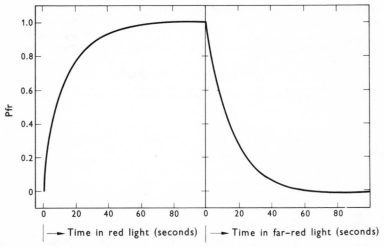

Fig. 3–1 Photoconversion of phytochrome showing first order kinetics. Light irradiance $3\mathrm{W\ m^{-2}}$.

reaction the rate of conversion of phytochrome from one form to the other is relatively insensitive to temperature.

Although the Pr form of phytochrome absorbs very little in the far-red region of the spectrum, the absorption spectra of Pr and Pfr overlap considerably in the red region. This means that under red light there is continuous interconversion of Pr and Pfr with a dynamic photostationary equilibrium. It is therefore impossible to obtain pure Pfr by irradiation of Pr, the absorption spectrum of red-irradiated phytochrome shown in Fig.

2–9 being that of a mixture of Pfr and Pr. The photostationary equilibria ($Pfr/P = \varphi$, where P is the total amount of photochrome) established by different wavelengths (Fig. 3–2) can be calculated from photoconversion kinetics and absorption data. In this way red light was shown to maintain 80% of phytochrome in the Pfr ($\varphi = 0.80$) form whereas far-red light maintains less than 5% Pfr ($\varphi < 0.05$).

Although the photoconversion of one form of phytochrome to the other is a simple first order reaction it is not a single step process. The actual photoreaction, which presumably involves an isomeric change of

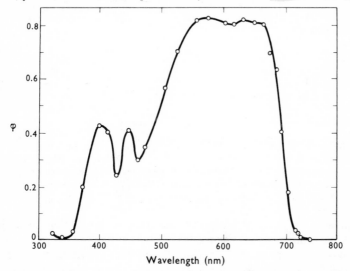

Fig. 3–2 Photostationary equilibrium ($\varphi = Pfr/P$) maintained by different wavelengths of light in mustard hypocotyls. (Redrawn from HANKE, J., HARTMANN, K. M. and MOHR, H. (1969). *Planta*, **86**, 235.)

the chromophore, is only the first step and is followed by a series of dark reactions before conversion is complete. The nature of these reactions and the intermediates between Pr and Pfr have been investigated by several techniques. Flash photolysis enables very rapid changes (less than a ms) in absorption to be measured. The data obtained using this technique show that after exposing Pr to a flash of red light several intermediate stages are passed through before Pfr is formed. A similar situation was observed for the reverse reaction. Low temperature studies have also provided useful information about phytochrome inter-mediates. At the temperature of liquid nitrogen ($-196°C$) conversion of Pr to Pfr and vice-versa is not possible, although intermediates are formed. Raising the temperature results in intermediates further along the pathways to Pfr and Pr respectively being formed. A scheme based on studies of phytochrome using both these techniques *in vivo* and *in vitro* is shown in Fig. 3–3. Data from studies on freeze-dried (dehydrated) phytochrome are consistent with this scheme. In the dehydrated state Pr can be photoconverted to lumi-R but the presence of water is necessary for the subsequent reactions since they involve changes in protein conformation. Although at room temperature the dark reactions are completed within a few seconds of the photoreaction, intermediates can

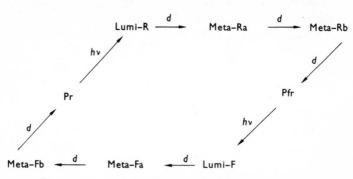

Fig. 3–3   Scheme proposed for the photoconversion of phytochrome. $hv=$ photoreaction; $d=$dark reaction. Lumi-R has an absorption peak at 698 nm. Lumi-F absorbs in the red region of the spectrum. Meta-Rb and meta-Fb are relatively weakly absorbing intermediates. (Redrawn from KENDRICK, R. E. and SPRUIT, C. J. P. (1973). *Plant Physiol.*, **52**, 327.)

be present in significant proportions. This is the case under mixed red/far-red light which excites both Pr and Pfr and so results in a continual cycling of the pigment. Under these conditions the dark reactions are important since they are the slowest point in the cycle. Therefore intermediates in the Pr → Pfr pathway accumulate, because these precede the slowest dark reaction (meta-Rb → Pfr). The intermediate meta-Rb is weakly absorbing compared to Pr or Pfr.

The nature of the chromophore isomerization is of great interest. The photoreaction at low temperature, e.g. Pr → lumi-R, is probably restricted to events within the chromophore and the dark reactions that occur at higher temperatures involve changes in both the chromophore and the protein. The protein changes are small relative to the overall molecular size of 120 000 daltons, but are no doubt of great significance at the 'active site' of the molecule. Immunological differences and differences in the ultra-violet absorption spectra have been reported. Another method of investigating conformational changes between Pr and Pfr is to react phytochrome with glutaraldehyde which specifically combines with free lysine in the protein. In Pr 13 lysine molecules are available for the reaction compared to only 11 in Pfr.

The chemical nature of the chromophore change has now been proposed (Fig. 2–10). A comparison of the absorption spectra of Pr, Pfr, and intermediates with those predicted for theoretical tetrapyrrole structures has provided information about the chromophore during conversion. Both Pr and Pfr correspond to extended forms of the tetrapyrrole molecule, whereas the intermediate meta-Rb corresponds to a tightly folded form (Fig. 3–4).

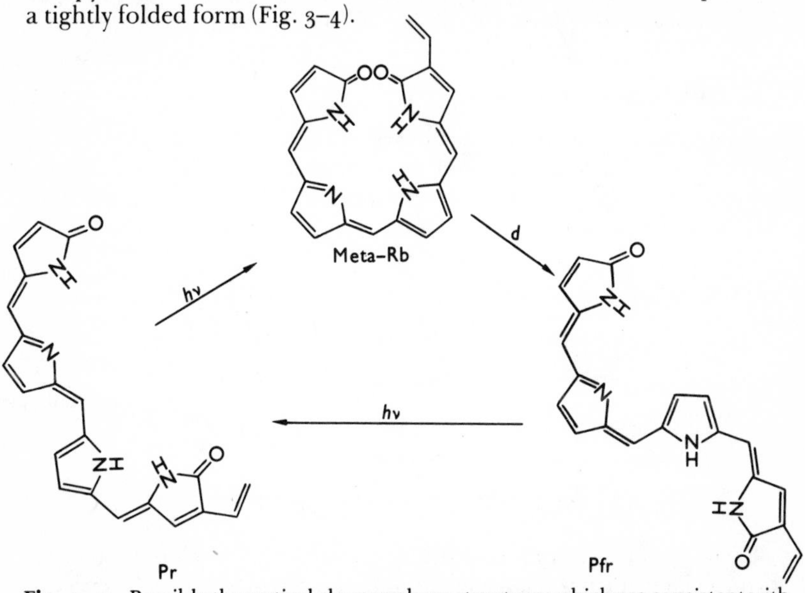

Fig. 3–4   Possible theoretical chromophore structures which are consistent with the absorption characteristics of Pr, Pfr and the intermediate meta-Rb. (Redrawn from BURKE, M. J., PRATT, D. C. and MOSCOWITZ, A. (1972). Reprinted with permission from *Biochemistry*, **11**, 4025. Copyright by the American Chemical Society.)

Phytochrome is not the only biological molecule that exhibits the phenomenon of photoreversibility. The visual pigment rhodopsin exhibits photoreversibility at low temperatures ($<-140°C$) with absorption maxima at 498 and 543 nm. Photoconversion of rhodopsin (Fig. 3–5) involves a cis-trans isomerization in the chromophore which is

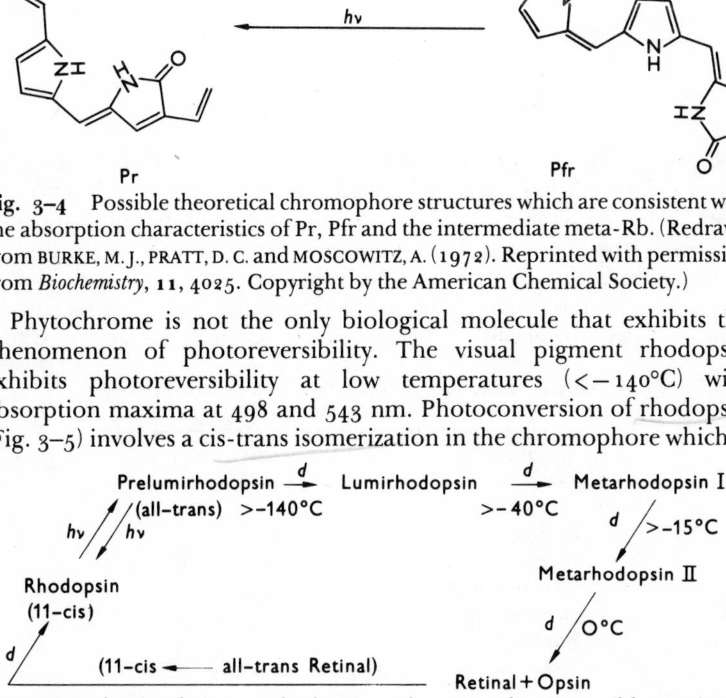

Fig. 3–5   The visual pigment rhodopsin undergoes a photoreversible reaction at low temperature. At higher temperatures the photoproduct of rhodopsin (prelumirhodopsin) undergoes a series of dark reactions to form retinol plus opsin.

followed by a series of dark reactions leading to a separation of chromophore (retinol) and protein (opsin). These end products reform rhodopsin by an enzymic reaction. The photoreversibility of rhodopsin can occur in the isolated chromophore whereas the phytochrome chromophore loses photoreversibility when detached from the protein.

3.2   Dark transformations

A great deal of work has been carried out on the dark reactions of phytochrome both *in vivo* and *in vitro*. Figure 3–6 summarizes the reactions that have been observed *in vivo* and Table 3.1 compares these reactions with those observed *in vitro*.

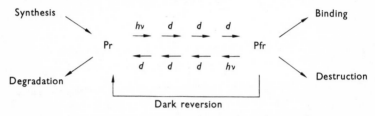

Fig. 3–6   Summary of the reactions of phytochrome *in vivo*.

Phytochrome is found in seedlings grown in the dark in the stable Pr form. The Pfr form is relatively unstable and *in vivo* following conversion from Pr by red light it undergoes a process called destruction or decay. Although this destruction is usually measured only as a loss of photoreversibility there is evidence that it involves a real loss of phytochrome molecules. The process requires oxygen and can be inhibited by metabolic inhibitors such as sodium azide. Under

Table 3.1   Reactions of phytochrome *in vitro* and *in vivo*

Reaction	In vitro	In vivo
Synthesis	−	+
Degradation	−	+
Dark reversion	+	+
Destruction (Decay)	−	+
Denaturation	+	−
Binding	+	+

continuous irradiation with red light the quantity of phytochrome decreases by this process to a low level. On falling below a critical value more phytochrome is synthesized *de novo* in the Pr form and an equilibrium is reached between synthesis and destruction. Pfr can undergo another reaction, the thermal dark reversion to Pr. This reaction was first predicted on the basis of physiological experiments (see 4.8). It takes place in all the dicotyledons so far investigated except for members of the Centrospermae. However, it is absent from most monocotyledons so far investigated. Figure 3–7 shows the changes in phytochrome levels which take place in the dark following conversion of Pr to Pfr by red light. The situation in sunflower (*Helianthus annuus*) which exhibits dark reversion is compared with that in love-lies-bleeding (*Amaranthus caudatus*) where only destruction takes place. Both destruction and dark reversion are temperature dependent.   — POETIC ELT!

The kinetics of Pfr destruction can be best studied in those species lacking dark reversion. In the dark, after a short exposure to red light, Pfr destruction follows first order kinetics with the rate of destruction at any given time (t) being proportional to the amount of Pfr remaining at that time.

$$\frac{dPfr}{dt} = -kPfr$$

Integrating this becomes:

$$Pfr = Pfr_0 e^{-kt}$$

Written in the logarithmic form:

$$\log_e \frac{Pfr}{Pfr_0} = -kt$$

Plotting the amount of Pfr on a logarithmic scale against time gives a straight line of gradient (k), the decay constant of Pfr. At 25°C, k is 0.037min^{-1} in *Amaranthus* seedlings corresponding to a half life ($t_{1/2}$) of about 20 minutes.

$$t_{1/2} = \frac{\log_e 2}{k}$$

Rates of destruction are rather slower in other species. For instance, in sunflower there is a half life of about 50 minutes at 25°C. Where dark reversion and destruction occur together analysis of the kinetics is more difficult. In most cases the results are not explained by assuming two simple first order reactions competing for Pfr.

The time course curves of phytochrome destruction under continuous irradiation are also exponential in form and in all cases there is a linear relationship between the logarithm of phytochrome concentration and time as shown in Fig. 3–8. The rate of disappearance of phytochrome at any time is proportional to the concentration of the unstable Pfr form at that time. Since the re-establishment of the photochemical equilibrium between Pr and Pfr is very rapid compared to destruction of Pfr, then

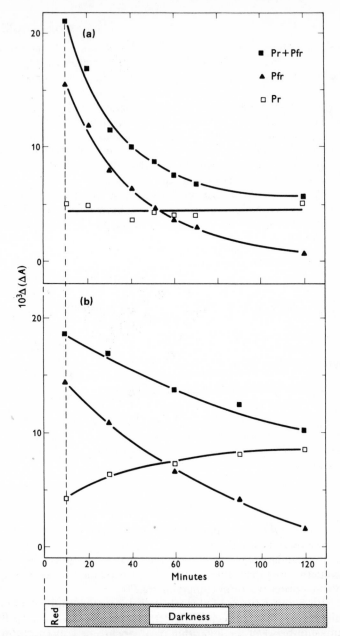

Fig. 3–7 (a) The dark transformations of phytochrome in *Amaranthus* seedlings after 10 minutes red light at 25°C. Demonstrates the process of destruction of Pfr and absence of dark reversion of Pfr to Pr. (Redrawn from KENDRICK, R. E. and FRANKLAND, B. (1969). *Planta*, **86**, 21.) (b) The dark transformations of phytochrome in sunflower hypocotyls after 10 minutes red light at 25°C. Demonstrates that Pfr destruction and dark reversion of Pfr to Pr occur simultaneously. (By courtesy of GLYNN, P. J. and FRANKLAND, B..)

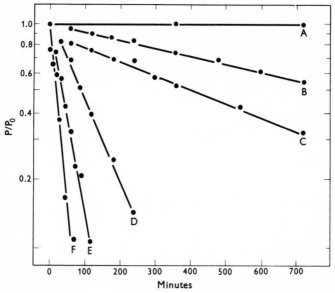

Fig. 3–8 Destruction of total phytochrome under continuous illumination at 25°C in *Amaranthus* seedlings. Total phytochrome as a proportion of that present initially (P/P$_0$) plotted on a logarithmic scale. (A) darkness. (B) far-red $\varphi = 0.022$. (C) far-red $\varphi = 0.045$. (D) blue light $\varphi = 0.22$. (E) mixed red/far-red light $\varphi = 0.49$. (F) decay of Pfr following 10 minutes red light $\varphi = 0.80$. (Redrawn from KENDRICK, R. E. and FRANKLAND, B (1968). *Planta*, 82, 317.)

under continuous irradiation the rate of decay of total phytochrome (P) will be given by the equation:

$$\frac{dP}{dt} = -k\varphi P$$

where k is the rate constant for destruction of Pfr and φ is the proportion of the phytochrome in the Pfr form (Pfr/P). The equation can be written in the logarithmic form thus:

$$\log_e \frac{P}{P_0} = -k\varphi t$$

When $\log_e$ P/P is plotted against time the gradient of the line is $k\varphi$, the rate constant for destruction of total phytochrome. Plotting this against φ gives a straight line with gradient equal to k (Fig. 3–9). Conversely φ can be calculated by determining the $k\varphi$ value graphically and dividing by k. Kinetics of this kind can be used in estimating Pfr/P ratios under various light treatments. This is especially useful with continuous far-red light where the amount of Pfr is very low and accurate measurement is not possible by direct spectrophotometry.

Pfr destruction in monocotyledonous seedlings, such as those of oats, follows zero order kinetics, i.e. there is a linear relationship between Pfr remaining and time. The rate of destruction is not dependent on the amount of Pfr present and is almost as rapid under continuous far-red light as under continuous red light.

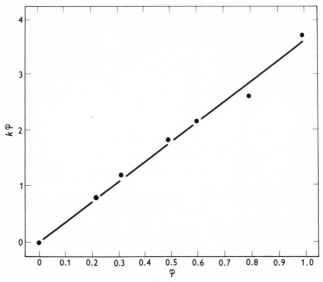

Fig. 3–9   Relationship between the decay constant of total phytochrome ($k\varphi$) and the photostationary equilibrium maintained (φ) in *Amaranthus* seedlings at 25°C. (Redrawn from KENDRICK, R. E. and FRANKLAND, B. (1968). *Planta*, **82**, 317.)

Another reaction that has been observed in seeds is the appearance of Pfr in darkness following irradiation with actinic far-red light. At first this was interpreted as a new phenomenon called 'inverse dark reversion' of Pr to Pfr. However, a better explanation is now available. Under conditions of dehydration found in seeds phototransformation of phytochrome is restricted. Far-red light converts Pfr to the intermediate lumi-F which can revert to Pfr in darkness if the forward reactions to Pr are restricted.

In early stages of phytochrome extraction from pea seedlings a process called rapid destruction has been observed. When the phytochrome is in the Pfr form it almost immediately loses photoreversibility. The active component in this reaction is a yet unidentified small molecular weight substance which has been called 'Pfr killer'. Phytochrome *in vitro* can easily be denatured by other compounds such as urea. Under these conditions Pfr forms a compound of much lower absorbance while still retaining photoreversibility. Denaturation can therefore be measured using the ratio of the absorbance change at the peak of Pfr (730 nm) to that at the peak of Pr (660 nm) following photoconversion. $\Delta A730/\Delta A660$ is equal to one for normal phytochrome but less than one for denatured phytochrome.

Dark reversion of Pfr to Pr has been observed in extracts of most species. Interestingly those species lacking the process *in vivo* such as the monocotyledons readily demonstrate the process *in vitro*. Low pH and a low oxidation-reduction potential enhance the process. It has been suggested that the presence of this reaction *in vivo* is correlated with the immediate environment of the phytochrome molecules. This is not the only change in properties that occurs upon extraction. The peak absorbance of Pfr shifts about 10 nm to a shorter wavelength. Also no Pfr destruction has been observed *in vitro*. It is possible that the different *in vivo* properties are due to a close association between phytochrome and other molecules such as membrane components.

Another reaction which has recently been observed by Marmé and co-workers is the preferential binding of the Pfr form of phytochrome to membrane fractions in the presence of magnesium ions at a suitable pH. Although the nature of this binding has not been characterized completely it is possible that it bears some significance to the localization of phytochrome in the cell and its physiological function.

3.3   Biosynthesis

The concentration of phytochrome in dry seeds is very low and usually not detectable. On addition of water the level of detectable phytochrome increases rapidly. This increase appears to be the result of the hydration of phytochrome molecules that are already present but do not show photoreversibility in the dehydrated state (Fig. 3–10). After a period the amount of phytochrome then increases again. This second increase has been shown by density labelling techniques to result from the *de novo* synthesis of phytochrome in the Pr form. This increase corresponds to the time of germination of the seeds and continues during seedling development until a plateau is reached. There is some evidence to support the idea that this does not represent a cessation of synthesis but rather a balance between Pr synthesis and Pr degradation. If total phytochrome is reduced by Pfr destruction, which must be a more rapid process than Pr degradation, a subsequent rise in phytochrome level can be observed in the dark. Under continuous irradiation with light a balance is eventually reached between destruction and synthesis (Fig. 3–11). It is therefore possible to conclude that a pool of phytochrome is present in light grown plants even though direct detection by spectroscopy is precluded. Extraction procedures have been successful in isolating phytochrome from the green leaves of several plants. Spectrophotometric investigations of white flower petals and the white regions of variegated leaves have also revealed low levels of phytochrome. One tissue from

34

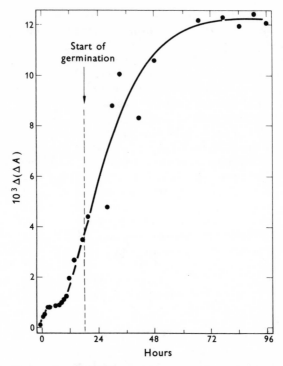

Fig. 3–10 Time course of phytochrome appearance in *Amaranthus* seed from the time of sowing in darkness at 25°C. (Redrawn from KENDRICK, R. E., SPRUIT, C. J. P. and FRANKLAND, B. (1969). *Planta*, **88**, 293.)

light-grown plants which is fairly high in phytochrome is cauliflower curd. In this case the phytochrome appears to be really photostable since on irradiation with red light there is no Pfr destruction but only dark reversion to Pr.

3.4 Phylogenetic aspects

Physiological responses suggest that phytochrome is present throughout the plant kingdom, from algae up to the higher land plants. There is some evidence for photoreversible systems in fungi and blue-green algae although the photoreceptors have not yet been identified. Positive identification of phytochrome by spectrophotometry or isolation has only been observed in a few lower plants. One striking difference appears to have been found between higher and lower plants. The absorption peaks of Pr and Pfr are shifted to shorter wavelengths in the green alga *Maesotaenium* and the bryophyte *Sphaerocarpus* (Table 3.2). *In vivo*

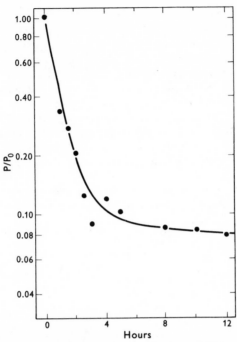

Fig. 3–11 Changes in total phytochrome in *Amaranthus* seedlings under continuous irradiation with red light at 25°C. Phytochrome as a proportion of that present initially (P/P_0) plotted on a logarithmic scale against time. Demonstrates that a balance is eventually reached between destruction and synthesis. (Redrawn from KENDRICK, R. E., SPRUIT, C. J. P. and FRANKLAND, B. (1969). *Planta*, **88**, 293.)

measurements of phytochrome in several gymnosperms (e.g. pine) similarly show absorption shifts to shorter wavelengths.

As mentioned in section 3.2 a relationship between *in vivo* dark

Table 3.2 Absorption characteristics of phytochrome *in vivo* and *in vitro*

	Group	Genus	Absorbance maxima (nm) Pr	Pfr
In vivo	Angiosperms	*Avena*	665	735
	Gymnosperms	*Pinus*	656	714
In vitro	Angiosperms	*Avena*	660	725
	Bryophytes	*Sphaerocarpus*	655	720
	Algae	*Maesotaenium*	649	710

reversion and taxonomy has been found in the species investigated. Dark reversion is absent in most monocotyledons and one order of dicotyledons, the Centrospermae. The latter group also shows other unique biochemical features.

Differences between phytochrome at the molecular level have been investigated using immunological techniques. Antisera against purified oat phytochrome shows some recognition for purified rye phytochrome. On the other hand oat antiserum shows very poor recognition of purified phytochrome from peas. It therefore appears that phytochrome has diversified during evolution, but clear connections are seen between related groups. Only when more complete data are available will a clear evolutionary record of phytochrome emerge. Obviously the photoreversible receptor pigment evolved at an early stage and its present ubiquitous distribution points to its success as a control system.

4 Phytochrome Controlled Responses

4.1 Photoresponses

Phytochrome is the photoreceptor involved in many developmental responses of plants to light. It is involved as a simple light detector and also, less directly, in measurement of light duration. The regulatory effects of light on plant growth and development are seen most dramatically at two stages in the life cycle of the plant. Firstly, at the stage of seed germination and seedling development, and secondly, at the stage of transition from the vegetative to the flowering phase (see 1.1).

All seedlings show developmental responses to light. These are associated with the transition from a seedling, dependent on food reserves and adapted for easy passage upwards through the soil, to a young plant with photosynthetic leaves. The morphological changes involved in this transition differ from species to species. For instance, in many dicotyledonous seedlings, such as sunflower (*Helianthus annuus*) or mustard (*Sinapis alba*), shoot extension in the dark is chiefly associated with the hypocotyl (Fig. 4–1). The upper part of the hypocotyl is in the form of a hook, the shoot apex and cotyledons pointing downwards and so being protected from damage as the seedling grows up through the soil. The effect of light is to inhibit hypocotyl elongation, to cause unfolding of the hypocotyl hook and to induce expansion of the cotyledons. Apart from the obvious synthesis of chlorophyll a variety of structural and physiological changes take place within the cotyledon as part of the transition from a storage organ to a photosynthetic leaf. In contrast to this type of seedling development some species, such as pea (*Pisum sativum*), show very little longitudinal extension of the hypocotyl, the cotyledons remaining below ground level (Fig. 4–1). Extension growth is the result of elongation of the first internodes. In seedlings grown in the dark there is again a terminal hook region protecting the shoot apex and very little leaf expansion. Light inhibits extension of the first internode, promotes extension of later internodes, causes hook unfolding and promotes expansion of the young leaves.

In graminaceous seedlings, such as oats (*Avena sativa*), there is no terminal hook but the young leaves are sheathed and protected by a modified leaf, the coleoptile (Fig. 4–1). The cotyledon remains within the seed and serves to absorb food materials from the endosperm. During the early stages of development of an oat seedling light inhibits extension of the first internode but promotes extension of the coleoptile. Although leaves are formed and grow in the dark they are tightly rolled. Light induces leaf unrolling as well as chlorophyll formation and differentiation of etioplasts to chloroplasts. Not all monocotyledons

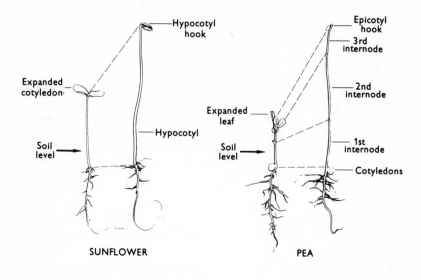

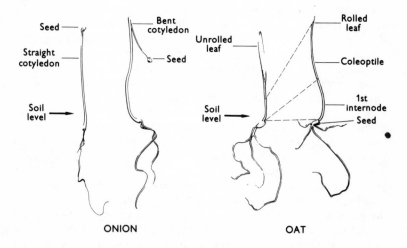

Fig. 4–1 Photomorphogenesis in seedlings. *Left:* grown in light. *Right:* grown in dark.

show this type of seedling development. In onion (*Allium cepa*), for instance, there is a narrow, elongated cotyledon, the tip of which is embedded in the endosperm of the seed (Fig. 4–1). Elongation of this cotyledon carries the seed above ground level. The cotyledon is bent at an acute angle and like the dicotyledonous hypocotyl shows a straightening response to light.

Not only does light have marked effects on seedling development but seed germination may also be affected by light. In many species germination is stimulated by light whereas in other species germination is inhibited by light. These effects of light may be seen both as changes in the proportion of seeds germinating and as changes in the time taken for seeds to germinate. The effect of light is frequently temperature dependent. For instance, lettuce (*Lactuca sativa*) seeds may germinate readily in both light and dark at 15°C, will germinate in neither light nor dark at 35°C but at 25°C show a low germination response in the dark with a marked positive response to light. In the negatively photoblastic seeds of love-lies-bleeding (*Amaranthus caudatus*) the temperature response is reversed with germination being favoured by high rather than low temperatures. A light requirement for germination is particularly characteristic of weed species such as charlock (*Sinapis arvensis*), goosefoot (*Chenopodium album*), dock (*Rumex obtusifolius*) and plantain (*Plantago major*). Such seeds tend not to germinate whilst buried in the soil but do so when exposed at the surface by cultivation.

The length of the day is one of the most important environmental variables determining the time of year when a particular species comes into flower (see 1.1). Long day plants (LDP) such as henbane (*Hyoscyamus niger*) flower when the days are longer than some critical length; SDP such as cocklebur (*Xanthium strumarium*) flower when the days are shorter than some critical length. Plants may show a 'quantitative' response to the length of the day. For instance, a LDP such as lettuce or charlock (Fig. 4–2) flowers most rapidly under long days but will eventually come into flower even under short days. In the case of cocklebur a single short day will induce flowering in plants which have been returned to non-inductive long days. This phenomenon of photoperiodism is found in relation to other aspects of plant development. For instance, the formation of dormant buds in many temperate-zone trees such as sycamore maple (*Acer pseudoplatanus*) is induced by short days. Although a period of high intensity light is necessary for most photoperiodic responses there is evidence that the most important factor is the length of the dark period. If the long dark period of a short day is interrupted by light a long day response will be produced (Fig. 1–1). Such light-break treatments are made use of commercially in maintaining glasshouse-grown *Chrysanthemum* plants, a short day plant, in the vegetative stage during autumn and winter. Plants can be brought into flower by terminating the light-break treatment. During the summer *Chrysanthemum*

40

Fig. 4–2 Photoperiodic control of flowering in charlock. Plants grown for 5 weeks under SD (*left*) or LD (*right*).

plants remain vegetative but can be brought into flower by covering them with black polythene in the morning and evening and so artificially shortening the day. It is now known that the length of the uninterrupted dark period is not the only controlling factor in photoperiodism. There is increasing evidence that the duration, intensity and quality of light during the light period can have effects on the flowering response. Also it has been shown in some species that during an artificially long dark period there is a rhythmic change in the response to a light break (see 4.13). Interruptions of the dark period after 24 and 48 hours tend to promote flowering whereas interruptions after 36 and 60 hours tend to inhibit flowering.

4.2 The classical red/far-red reversible reaction

Phytochrome is assumed to be the photoreceptor where induction of a response by a short irradiation with red light can be reversed by a subsequent irradiation with far-red light. This is the classical red/far-red reversible situation although phytochrome may be involved in other kinds of photoresponses. Action spectra show a peak for induction at 660 nm and a peak for reversion of this at 730 nm. These action peaks correspond to the peaks of absorption of Pr and Pfr respectively (see 1.2).

Red/far-red reversible responses have been observed in many aspects of seedling photomorphogenesis, including growth, structural and biosynthetic changes. Red/far-red reversibility is a feature of light stimulated seed germination in several species and of the light break in photoperiodic effects on flowering (see 1.1). Red/far-red reversible responses have also been found in lower plants such as ferns (e.g. spore germination in *Osmunda*), liverworts (e.g. thallus growth in *Marchantia*) and algae (e.g. chloroplast orientation in *Mougeotia*).

4.3   Rate of response

A short period of red light induces a response which takes place in the dark some time after the exposure to light. In the case of red light induced growth of apical buds in pea seedlings there is a lag period of about 4 hours before there is a measurable difference between irradiated seedlings and those maintained continuously in the dark (Fig. 4–3). In the case of light induced germination of lettuce seeds the response time at 25°C is about 12 hours. Some responses, however, are very rapid, an example being the nyctinastic closing movement of the leaflets of the

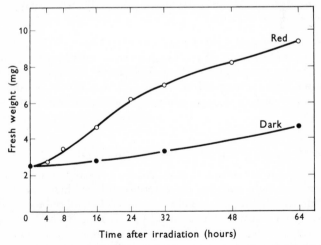

Fig. 4–3   Growth of apical buds of dark-grown pea seedlings at various times following red irradiation. (Redrawn from FURUYA, M. and THOMAS, R. G. (1964). *Plant Physiol.* **39**, 634.)

sensitive plant (*Mimosa pudica*). Leaves transferred from light to dark show a closing response after 5 minutes and complete closing within 30 minutes. The response is prevented by a short exposure to far-red light before the leaves are transferred to the dark.

4.4   Relationship between responses and incident light energy

Responses can be induced by very low amounts of light energy. Figure 4–4 shows the dose-response relationship for red light inhibition of first internode extension in oat seedlings. The response is proportional to the

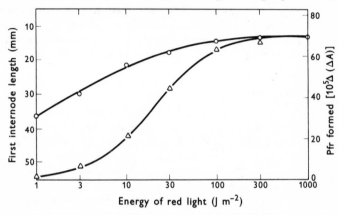

Fig. 4–4   Dose-response relationship for red light inhibition of first internode extension (o) and Pfr formation (Δ) in oat seedlings. (Redrawn from LOERCHER, L. (1966). *Plant Physiol.* **41**, 932.)

logarithm of incident light energy and is saturated at 100 J m^{-2}. This quantity of red light would be contained in about 2 seconds of normal sunlight! Also note that the oat first internode will show a small but significant response to 0.1 J m^{-2}. This amount of light is sufficient to saturate very sensitive systems such as the hook opening response in bean seedlings!

4.5   Is Pfr the physiologically active form of phytochrome?

Red light may have promoting or inhibiting effects. For instance, in pea seedlings red light inhibits the extension of the first internode but promotes leaf expansion. It is possible to think of all red light effects as positive. Light inhibits cell extension by promoting cell maturation.

The primary effect of red light is to convert Pr to Pfr. Is the inhibition of internode extension due to the presence of Pfr or to the absence of Pr? In other words, is Pfr the physiologically active form of phytochrome? The data in Fig. 4–4 can be re-plotted to show that there is a simple linear relationship between the response and the logarithm of the amount of Pfr (Fig. 4–5). A change in the Pfr/P ratio from say, 0.01 to 0.05 produces a large response. The concurrent change in Pr/P ratio is from 0.99 to 0.95. If Pr were the active form it is difficult to see why a small relative change in Pr should produce such a large response. On the basis of circumstantial

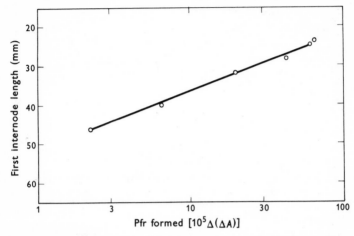

Fig. 4–5   Red light inhibition of first internode extension in oat seedlings as a function of Pfr. Data as for Fig. 4–4. Pfr plotted on a logarithmic scale. (Redrawn from LOECHER, L. (1966). *Plant Physiol.* **41**, 932.)

evidence of this kind Pfr is taken to be the physiologically active form of phytochrome.

4.6   Relationship between responses and Pfr/P ratio

Figure 4–5 shows a simple relationship between responses and Pfr/P ratio in oat seedlings. Figure 4–6 gives data for light inhibition of extension of pea stem sections. Stem sections were exposed for 15 minutes to various mixtures of red and far-red light establishing various photostationary states. Here again the response is related to Pfr/P ratio although values of 0.5 and above are saturating. These are *graded* responses to Pfr in contrast to an ungraded or *threshold* response as shown by Mohr and coworkers in the light inhibition of lipoxygenase formation in mustard cotyledons. Light giving a Pfr/P ratio greater than 0.01 completely inhibits lipoxygenase formation whereas light giving a Pfr/P ratio less than 0.01 has no effect on the level of enzyme activity.

If the experiment with pea stem sections, described above, is repeated with sections from seedlings that were irradiated 9 hours previously with red light, the data shown in Fig. 4–7 are obtained. Far-red light, which establishes about 5% of phytochrome in the Pfr form, now promotes

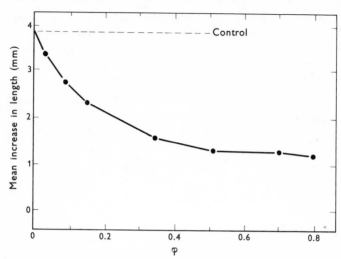

Fig. 4–6 Elongation of 5 mm stem sections from dark-grown pea seedlings following irradiation for 15 minutes with various mixtures of red and far-red light establishing different Pfr/P ratios (φ). (Redrawn from HILLMAN, W. S. (1965). *Physiol. Plant*, **18**, 346.)

elongation, presumably by reversing Pfr to Pr. There is no effect of light producing 30% Pfr suggesting that the tissue already contains 30% Pfr. However, spectrophotometric measurements show that this is not so. Pfr is unstable and in the dark either undergoes destruction or dark reversion

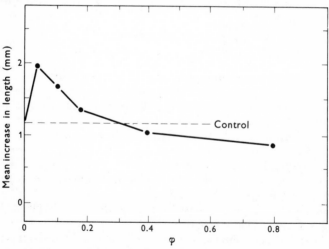

Fig. 4–7 Experiment as in Fig. 4–6, but stem sections from pea seedlings irradiated 9 hours previously with red light. (Redrawn from HILLMAN, W. S. (1965). *Physiol. Plant*, **18**, 346.)

to Pr (see 3.2). This situation, where spectrophotometry fails to confirm the presence of Pfr which has been deduced from physiological responses to light, has been described by Hillman as a 'phytochrome paradox'. It is generally true that it is difficult to detect phytochrome in light-grown plants, unless a very sensitive, specialized spectrophotometer is used, despite the fact that red/far-red reversible responses may be observed in such plants. One explanation is that there is a small 'pool' of physiologically active phytochrome distinct from the 'bulk', photometrically detectable, phytochrome. It has to be postulated that the Pfr of the 'active' phytochrome is relatively stable and persists in the dark for a relatively long period. On the other hand it has been proposed by Kendrick and Spruit that all phytochrome is active and controls specific responses by virtue of its localization in particular molecular environments within the cell, where its properties may be significantly different. Spectrophotometric studies may only indicate the average properties of a large number of small, physiologically active pools of phytochrome controlling different responses and having different properties.

4.7   Transmissible phytochrome effects

Plant hormones, such as auxins and gibberellins, are involved in the internal regulation of plant growth and it seems probable that some phytochrome controlled responses involve changes in the level of such hormones. A red/far-red reversible response may take place in one part of the plant following irradiation of another part. For instance, in bean seedlings irradiation of the hypocotyl hook region with red light can induce expansion of the young leaves. Clearly there is transmission of a stimulus. Some transmissible phytochrome effects may be mediated by hormones but in this particular example the rate of transmission is too rapid for this to be a likely explanation. In the flowering response there is transmission of a stimulus over a long distance from the leaves where daylength is perceived to the growing apex where floral primordia are formed. It has been proposed that the flowering stimulus is a hormone, florigen, although it has not yet been isolated and identified.

4.8   Escape from reversibility: time course of Pfr action

If there is a sufficiently long dark period between irradiation with red light and a subsequent far-red irradiation there is loss of reversibility. Figure 4–8 presents data for red light inhibition of coleoptile growth in intact rice seedlings and shows the progressive loss of reversibility by far-red as the length of the intervening dark period is increased. There is 50% loss of reversibility after 8 hours. Such 'escape curves' are of great interest in that they indicate the time course of Pfr action. During the dark period

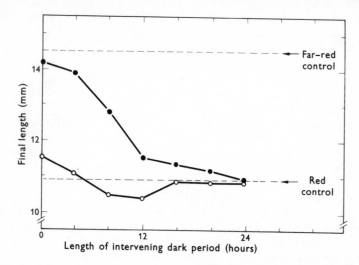

Fig. 4–8 Effect of intervening dark period between red and far-red light on escape of red light induction of inhibition of rice coleoptile elongation from far-red reversal at 27°C. ●——● Dark grown intact coleoptiles exposed to red light for 3 minutes and returned to darkness for various periods before exposure to far-red light for 3 minutes. o——o As above, but a second red irradiation given immediately after the far-red light. Final coleoptile length measured 2 days after first red irradiation. (Redrawn from PJON, C. J. and FURUYA, M. (1967). *Plant and Cell Physiol.* **8**, 709.)

Pfr is active and although it can be reversed to Pr by far-red light its action cannot be so reversed. The rate of escape from far-red reversibility is usually temperature dependent. For instance, in lettuce seed germination the time for 50% loss of reversibility is 9 hours at 20°C and 5 hours at 25°C.

Figure 4–9 shows fairly rapid escape from far-red reversibility of inhibition of flowering in the SDP cocklebur by a red light interruption of the dark period. This indicates rapid *potentiation* or *induction* of the response although the response itself is not rapid. It is three days before a change from a vegetative to a flowering apex is visible in plants exposed to a long night which was not interrupted by light.

In photoperiodic control of flowering, red light given at the end of the light period has no effect whereas red light given in the middle of the dark period inhibits flowering in SDP and promotes flowering in LDP. Red light can only act by converting Pr to Pfr and therefore phytochrome in the Pr form must have appeared during the dark period. This is taken as physiological evidence for the non-photochemical reversion of Pfr to Pr (see 3.2).

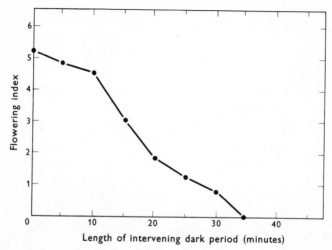

Fig. 4–9 Effect of various durations of darkness between red and far-red irradiation on the promotion of flowering of cocklebur. (Redrawn from DOWNS, R. J. (1956). *Plant Physiol.* **31, 279**.)

4.9 Apparent incomplete reversal by far-red

In seedling growth responses it is often found that far-red light does not completely reverse the effect of red light. In such cases far-red light alone usually produces the same response as red light followed by far-red light. The small but significant response is due to the small proportion (about 5%) of Pfr established by far-red light.

Incomplete reversal can also be due to rapid potentiation with some Pfr action taking place during the red irradiation period. In this case red light followed by far-red light will produce a larger response than far-red light alone.

4.10 Requirement for repeated irradiation with red light

Prolonged irradiation with red light may be necessary for the induction of certain responses. In the case of light inhibition of cucumber hypocotyl elongation the response can be induced by repeated short exposures to red light over a period of time. The red irradiations are far-red reversible. There is clearly a need for Pfr action over a long period of time, possibly due to the supply of some co-effector limiting the response. Alternatively there may be a need for the accumulation of some product of Pfr action. Responses of this kind indicate that, following each red irradiation, there must be a fall in Pfr level (e.g. by dark reversion to Pr, or by Pfr destruction followed by Pr synthesis).

4.11   Far-red inhibition of seed germination

Some seeds, for instance those of certain varieties of tomato, germinate in the dark but can be inhibited by a short exposure to far-red light. This effect of far-red light can be reversed by red light. This suggests that phytochrome must be present as Pfr in the seed in the dark.

Some seeds, such as those of *Amaranthus* and some varieties of lettuce, which germinate in the dark, require prolonged far-red illumination to inhibit germination. Dormancy can be induced by prolonged far-red illumination but this can be broken by a single short red irradiation. This situation suggests that Pfr is appearing in the seed in the dark over a prolonged period of time. A simple explanation is possible if it is assumed that in the seed phytochrome is not in a fully hydrated state, so that the dark steps in the conversion of Pr to Pfr and Pfr to Pr are slowed down (see 3.1). On transfer of seeds from far-red light to dark Pfr may arise from intermediates.

In seed germination prolonged far-red irradiation acts in the opposite way to short red irradiation. In contrast, in seedling photomorphogenesis prolonged far-red irradiation acts in the same way as short red irradiation despite the fact that short far-red reverses the effect of short red (see 4.14).

4.12   Phytochrome action in light-grown plants

It has already been mentioned that red/far-red reversibility can be demonstrated in relation to light interruption of the dark period in experiments on the photoperiodic control of flowering. This clearly shows that phytochrome is involved as a light detector in normal light-grown, green plants as well as in dark-grown seedlings and seeds sown in the dark.

Phytochrome can also be shown to be involved in the regulation of growth in light-grown plants. For instance, the elongation of stem internodes in bean plants grown under 8 hour days can be increased by a factor of 4 by irradiating the plants with 5 minutes far-red light at the end of the light period. The type of growth obtained depends on whether or not Pfr is present during the dark period.

4.13   Phytochrome and biological time measurement

Photoperiodic responses must involve time measurement, particularly measurement of the length of the dark period. At first it was suggested that the dark reversion of Pfr to Pr might provide the basis of an 'hour glass' type of clock. However, the process is not temperature independent and for this and other reasons the idea was rejected. The existence of a number of rhythmic phenomena suggests that there is an oscillating type of 'biological clock'. Rhythms have been observed in the 'sleep movements' of leaves and in biochemical parameters such as carbon dioxide output, activities of various enzymes and levels of adenosine triphosphate. Such rhythms are described as *endogenous*, since they continue after transfer from normal light/dark cycles to constant environmental conditions, and *circadian*, since they have a periodicity of approximately 24 hours. Phytochrome has been implicated in the initiation and the phase-setting of such rhythms. In photoperiodism phytochrome may be involved in starting an oscillating system which measures the dark period. However, the clock may be independent of phytochrome but control changes in sensitivity to Pfr during the dark period (see 4.1).

4.14   The high irradiance reaction

Plants are not normally exposed to short periods of red and far-red light. Normally they are exposed to prolonged, high intensity white light from the sun. Although a short exposure to red light may induce marked changes in seedling development it does not transform a seedling into a normal looking green plant. Prolonged illumination is necessary for this.

Figure 4–10 shows that for prolonged or continuous illumination there are peaks of action in the blue and far-red regions of the spectrum. Prolonged far-red irradiation, unlike short far-red irradiation, acts in the same way as short red and is, in fact, more effective than prolonged red irradiation. Mohr has described this as the high energy reaction or *high*

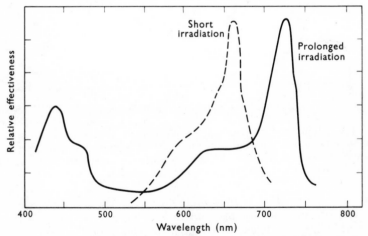

Fig. 4–10 Comparison of action spectra for prolonged and short irradiations. The two curves are plotted on different scales (short red is, of course, less effective than prolonged red irradiation). (Redrawn from MOHR, H. (1959). *Planta*, 53, 219.)

irradiance reaction. Originally it was suggested that there was a photoreceptor with absorption peaks in the blue and far-red regions of the spectrum but there are other explanations.

Hartmann investigated the high irradiance reaction using dual wavelength experiments. The systems used were the photoinhibition of lettuce hypocotyl extension and the photoinhibition of lettuce germination. Both responses require prolonged illumination and there is increasing response with increasing irradiance. In both cases there is a sharp peak of action at about 720 nm. It was found that the action of 720 nm light could be nullified if mixed with otherwise inactive light, either from the red or from the infra-red side of 720 nm. There was also an enhancement effect where action could be produced by an appropriate mixture of two otherwise inactive wavelengths (Fig. 4–11). Effects of this kind are difficult to interpret in terms of a photoreceptor with a peak of absorption at 720 nm. They can be interpreted in terms of phytochrome if

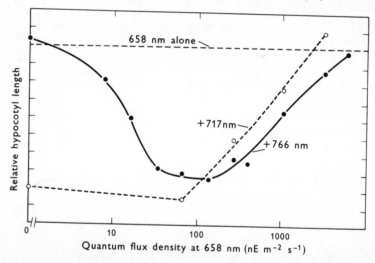

Fig. 4–11 Inhibition of hypocotyl elongation in lettuce seedlings under continuous simultaneous irradiation with different quantum flux densities at 658 nm and a constant high flux background of either 717 nm or 766 nm. (Redrawn from HARTMANN, K. M. (1966). *Photochem. Photobiol.* 5, 349.)

one assumes that there is an optimal Pfr/P ratio for action. Maximum effect is obtained with a light mixture giving a Pfr/P ratio of 0.06, whereas at higher or lower Pfr/P ratio there is less action.

Why does response increase with increase in irradiance? One must postulate that action depends on both the photostationary state, which is energy independent, and the photostationary flux, which is energy dependent. Although an increase in irradiance does not change the Pfr/P ratio it does increase the rate of interconversion of Pr and Pfr and this may be significant in explaining high irradiance effects (see 5.1).

Although phytochrome absorbs in the blue region of the spectrum it has been argued that blue light acts through a different photoreceptor. Evidence for this comes from observed qualitative differences in the effects of prolonged blue and prolonged far-red light. For instance, gherkin seedlings transferred from the dark to blue light show a reduced rate of hypocotyl extension after a lag of only 30 seconds whereas the lag period for far-red light is 40 minutes. Also, lettuce seedlings at one stage of development respond to both blue and far-red whereas at a later stage they respond only to blue. In the extensively studied phototropism of grass coleoptiles it is the blue region of the spectrum which induces a growth curvature towards a unilateral source of light. Action spectra suggest that the photoreceptor is a yellow pigment such as a flavoprotein or a carotenoid. However, red light acting through phytochrome can modify the phototropic response of coleoptiles to a subsequent exposure to unilateral blue light.

4.15 Phytochrome and plants in their natural environment

Figure 4–12 shows the spectral distribution of sunlight. It contains rather more red than far-red light and would be expected to maintain a

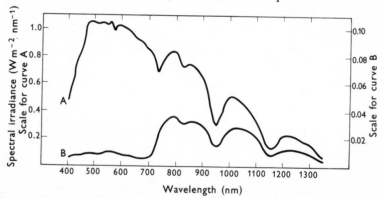

Fig. 4–12 Spectral distribution of light in the open at noon on a sunny day in July (A) compared to that of light within a woodland (B). (By courtesy of LETENDRE, R. J. and FRANKLAND, B.)

fairly high proportion of phytochrome in the Pfr form. Plants under natural conditions are either in the dark (low Pfr) or exposed to sunlight (high Pfr). But what is the function of far-red reversibility? It should be remembered that green leaves are very effective absorbers of red light. In fact, photoreversibility can be demonstrated in light-requiring seeds by briefly exposing them to direct sunlight followed by exposure to sunlight filtered through a green leaf. Therefore, phytochrome within a leaf, particularly a leaf on a plant in dense vegetation, will be exposed to light with a relatively high far-red/red ratio. This may well be of ecological significance and perhaps give a clue to the selection pressures operating in the evolution of phytochrome as a light detector.

5 Mode of Action of Phytochrome

5.1 Analysis of Pfr action

Pfr is regarded as the physiologically active form of phytochrome (see 4.5). Conversion of Pr to Pfr by light will produce a particular response depending on the localization of the phytochrome and the state of differentiation of the responding cell or cells. This heterogeneity of phytochrome distribution within and amongst cells may well explain many of the failures to find a correlation between photoresponse and photometrically detectable Pfr.

In the case of the high irradiance reaction dependence of response on Pfr is rather complex with high and low Pfr/P ratios being less effective than intermediate ratios. There is also irradiance dependence suggesting that the rate of conversion of Pr to Pfr is important in producing the response. One suggestion is that there is an excited species of Pfr (Pfr*) which is more active than Pfr in the ground state. Under continuous illumination the proportion of phytochrome as Pfr* will increase with increase in light intensity. Consideration must be given to the possibility that Pfr* is one of the intermediates described in section 3.1.

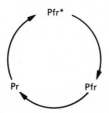

Pfr action has been studied in two main experimental situations. Firstly, conversion of Pr to Pfr by a single short pulse of red light, the induced response being followed in the subsequent dark period. Secondly, by continuous irradiation with far-red light maintaining a low but constant (steady state) Pfr/P ratio. The first method allows the time course of escape from far-red reversion to be followed. This gives an indication of the time course of Pfr action. Application of metabolic inhibitors before and after irradiation can be used to give some indication of the prerequisites for Pfr action and the nature of Pfr action. However, it is often difficult to separate effects on Pfr action from effects on later processes leading up to the response.

It appears likely that there are several steps between the initial action of

Pfr and the final response. Very little is known about this sequence of events.

Attention has been focused on phytochrome controlled biochemical changes since here there is a greater chance of tracing back the sequence of biochemical steps to the point of initial Pfr action. Red/far-red reversibility has been observed in relation to the synthesis of pigments such as chlorophyll, carotenoids and flavonoids (e.g. anthocyanins). There is also phytochrome control of the breakdown of carbohydrates and other food reserves in seedlings. For instance, starch stored in the leaves of maize seedlings is mobilized more rapidly in red irradiated seedlings as compared to seedlings maintained continuously in the dark.

The fact that phytochrome is a protein led to the idea that Pfr was the active form of an enzyme. Light induced changes in the chromophore lead to changes in the conformation of the protein which could increase the accessibility of certain active groups (e.g. SH groups) which then allows binding with the substrate.

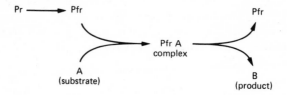

However, there is no indication of the nature of the substrate nor is there any convincing experimental evidence that phytochrome is an enzyme. Nevertheless, the general idea that conversion of Pr to Pfr involves a conformational change in the protein followed by binding to some co-effector molecule has been retained.

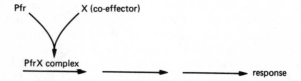

A theoretical model, recently proposed by Schäfer, attempts to explain the difference between the high irradiance reaction and the low energy, inductive reaction in terms of changes in the PfrX complex. PfrX' is the form present in the dark following a brief irradiation whereas PfrX is the form which will tend to accumulate under continuous, high irradiance conditions. The model in its complete form takes account of both Pr

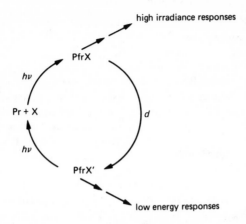

synthesis and Pfr destruction. It can also be used to explain why the high irradiance peak of action is in the far-red region of the spectrum.

5.2 Phytochrome and gene expression

It is generally accepted that most living cells of a particular plant contain all the genetic information (in the form of DNA, deoxyribonucleic acid) characteristic of that plant. Differences amongst cells must arise from differential gene activity, genes being 'turned off' and 'turned on' during development. Many of the responses controlled by phytochrome involve changes in the direction of differentiation and hence changes in the pattern of gene activity. This led Mohr in 1966 to suggest that phytochrome might act by controlling gene activity. He proposed that 'primary' differentiation of a cell produced a certain pattern of gene activity with some of the inactive genes having the potential for being activated by Pfr and some of the active genes being capable of being repressed by Pfr. The changes resulting from Pfr action he described as 'secondary' differentiation.

The sequence of amino acids in a protein is coded for by the sequence of nucleotide bases in the DNA, the flow of information being summarized as follows:

TRANSCRIPTION   TRANSLATION

DNA (gene) ⟶ messenger RNA ⟶ protein

nucleus   cytoplasm
(chromosomes)   (ribosomes)

Metabolic inhibitors have been used to explore the point of action of Pfr although the results of such experiments have to be interpreted with caution. For instance, actinomycin-D, an inhibitor of transcription (i.e.

DNA-dependent RNA synthesis), will inhibit certain phytochrome controlled responses. This may only indicate that Pfr action or a later process in the response is dependent on RNA synthesis rather than that Pfr acts by stimulating RNA synthesis.

Increased ribosomal RNA synthesis has sometimes been observed in association with phytochrome controlled growth responses although this may be a consequence of increased growth rather than a direct result of Pfr action. The appearance of polyribosomes in dark-grown leaves exposed to light can be taken as suggesting messenger RNA synthesis although, again, there is no strong evidence that this is a very direct result of Pfr action.

Some phytochrome controlled responses are relatively insensitive to RNA synthesis inhibitors but can be blocked by inhibitors of protein synthesis such as puromycin or cycloheximide. This could suggest that phytochrome is acting at the level of translation (i.e. in mRNA guided synthesis of proteins on the ribosomes) rather than at the level of transcription.

5.3   Phytochrome and enzyme synthesis

Whatever the mode of action of Pfr there is little doubt that it often leads to changes in the pattern of enzymes present. For instance, seedling photomorphogenesis is associated with the appearance of enzymes necessary for photosynthesis. NADP-dependent glyceraldehyde-3-phosphate dehydrogenase, an enzyme associated with leaf chloroplasts, was one of the first enzymes in which changes in activity were shown to be red/far-red reversible.

An enzyme which has been extensively studied is phenylalanine ammonia lyase. This enzyme catalyses the conversion of the amino acid phenylalanine to cinnamic acid and thus redirects metabolism away from protein synthesis and towards synthesis of various phenolic compounds such as the red anthocyanin pigments and other flavonoids. Other substances eventually formed by this pathway include lignin, a constituent of secondary cell walls, and coumarin (responsible for the smell of new mown hay).

The enzyme is present in very low concentrations in mustard seedlings grown in the dark but can be greatly increased by exposure to light. The time course of changes in enzyme activity on transfer from dark to far-red light is shown in Fig. 5–1. There is a lag phase of about one hour and then

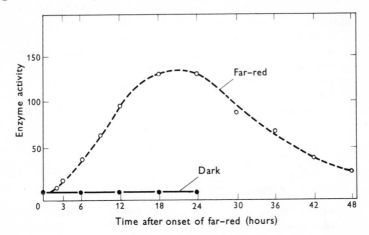

Fig. 5–1　Kinetics of appearance of the enzyme phenylalanine ammonia lyase in cotyledons of mustard seedlings exposed to continuous far-red light. Enzyme activity expressed as pmoles cinnamic acid formed per minute per pair of cotyledons. (Redrawn from DITTES, L., RISSLAND, I. and MOHR, H. (1971). *Naturforsch.* **26b**, 1175.)

a rise in enzyme activity, followed by a fall in activity after about 20 hours. Such kinetics can be interpreted in terms of the following sequence of events: (i) induction of enzyme synthesis, (ii) inactivation of enzyme, (iii) repression of enzyme synthesis. Inactivation and repression are initiated by the build-up of end products of enzyme action such as cinnamic acid and p-coumaric acid.

The increase in enzyme activity can be blocked by protein synthesis inhibitors but this is not rigorous proof that there is an actual increase in the number of enzyme molecules. There is always doubt as to the specificity of action of a given inhibitor and, in any case, an inhibitory effect only indicates that protein synthesis is a pre-requisite for the process. Density labelling experiments do provide some evidence that *de novo* enzyme synthesis is taking place. Such experiments involve growing seedlings in heavy water (D_2O, deuterium oxide), extracting the enzyme and determining its density by appropriate centrifugation techniques. During hydrolysis of storage protein, deuterium is incorporated into the amino acids formed. These 'heavy' amino acids will become incorporated into newly synthesized protein and thus increase its density relative to protein synthesized in seedlings grown in ordinary water. Even the interpretation of this type of experiment has been criticized by some scientists who claim there is still no convincing proof that Pfr action leads to synthesis of the phenylalanine ammonia lyase enzyme. They suggest the effect of Pfr could be on activation of previously synthesized enzyme molecules. However, there is agreement that Pfr action results in the synthesis of some enzymes, e.g. ascorbic acid oxidase. In most cases Pfr simply alters the rate of synthesis of enzymes already present in seedlings grown in the dark. This is difficult to explain on the basis of a simple model of induction of enzyme synthesis.

Some enzymes, such as lipoxygenase, are formed in the dark in mustard seedlings but are completely repressed by Pfr. Other enzymes such as catalase and isocitrate lyase appear to be unaffected by light.

5·4 Phytochrome and permeability

Although changes in gene activity are an eventual consequence of Pfr action there is no evidence that this is the *primary* mode of action of phytochrome. Some photoresponses are so rapid that it is difficult to believe that gene action and protein synthesis are involved. Such a rapid response is the phytochrome controlled dark closure or folding of the leaflets of *Mimosa* and *Albizzia*. The response involves differential changes in turgor in the cells of the pulvinus at the base of each leaflet. The turgor change is associated with the movement of potassium and other ions into ventral cells and out of dorsal cells. This has led to the suggestion that the primary action of phytochrome is on membrane permeability.

Further evidence of membrane changes following Pfr action came from Tanada's discovery that excised barley and mung bean root tips exposed to red light would stick to a negatively charged glass surface. This peculiar phenomenon of light induced adhesion was found to be red/far-red reversible. It was suggested that the apical part of the root segment became electro-positive relative to the basal part in response to the formation of Pfr. Measurements of the potential difference between apex and base confirmed this. The changes in electrical potential were only about one millivolt but developed within 30 seconds of the light treatment. Rapid phytochrome controlled changes in electric potential have also been measured in the coleoptiles of oat seedlings. These electrical changes are consistent with a phytochrome induced efflux of ions. Jaffe's observation of red light induced changes in acetylcholine levels in plant tissues suggested similarities between Pfr action and the generation of an action potential in the nerve fibres of animals. However, there is no evidence for the involvement of acetylcholine in other phytochrome controlled responses.

5.5   Intracellular localization of phytochrome

Elegant experiments carried out by Haupt between 1960 and 1970 on the photocontrol of chloroplast movement (Fig. 5–2) in the alga *Mougeotia* have established that phytochrome is located not in the chloroplast but in or near the plasma membrane. Using microbeams and polarized light it was shown that Pr must have its axis of maximum absorption in the plane

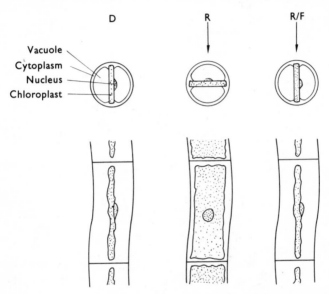

Fig. 5–2 Phytochrome control of chloroplast orientation in cells of the alga *Mougeotia*. Cells shown in cross-section and surface view. D, dark. R, orientation of chloroplast 30 minutes after 1 minute irradiation with red light. R/F, orientation of chloroplast 30 minutes after irradiation with 1 minute red light immediately followed by 1 minute far-red light. (Redrawn from HAUPT, W. (1970). *Physiol. vég.* **8**, 551, Editions Gauthier–Villars, Paris.)

of the cell surface whereas Pfr has its axis at right angles to the cell surface. There is *dichroic* orientation of the phytochrome molecules. This means that unilateral red light will tend to transform Pr to Pfr more readily at the front and back of the cell compared to the sides. Assuming that the flat chloroplast moves away from regions of the cytoplasm with high Pfr levels then an explanation is provided for the orientation of the chloroplast at right angles to the incident light. This is illustrated diagrammatically in Fig. 5–3.

Fern sporelings will show growth curvatures related to the plane of vibration of the electrical vector of polarized light. Growth takes place at the tip of the apical cell and the response to polarized light can be

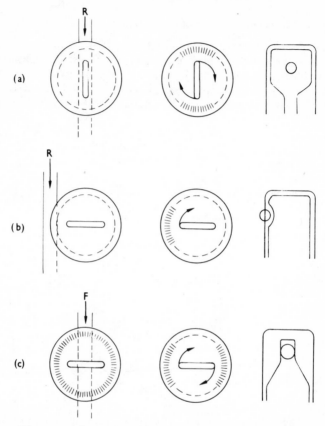

Fig. 5–3 Diagrammatic representation of three experiments involving irradiation of *Mougeotia* cells with red (**R**) or far-red (**F**) microbeams. In (**a**) and (**b**) cells previously in dark; in (**c**) cells previously in light. *Left:* cross-sections showing starting condition and position of microbeam. *Centre:* cross-sections showing conditions after phytochrome photoconversion, curved arrows show direction of movement of chloroplast. *Right:* surface view with position of microbeam and response of chloroplast. Tangentially oriented dashes indicate Pr; radial dashes indicate Pfr. Note how the chloroplast moves away from Pfr. (Redrawn from HAUPT, W. (1970). *Physiol. veg.* **8**, 551, Editions Gauthier–Villars, Paris.)

explained in terms of phytochrome molecules orientated in the peripheral cytoplasm. Such experiments do not rule out the possibility of phytochrome being located in other parts of the cell. For instance, red/far-red reversible changes have been observed in isolated mitochondria and plastids. Both changes in ultrastructure and rapid release of the plant hormone gibberellin have been observed in etioplasts isolated from cereal leaves.

Attempts are currently being made to locate phytochrome within the cell by microspectrophotometry and immunocytochemical methods. Purified phytochrome can be injected into an animal to produce a highly specific antiserum which can then be used to locate phytochrome in tissue sections. Preliminary results show that phytochrome is present in the cytoplasm as well as being associated with cell organelles.

5.6 Phytochrome in membranes

Attempts have been made to detect phytochrome associated with cell organelles from fractionated tissue homogenates. During normal methods of phytochrome extraction a small proportion of the total phytochrome does appear in the 'pellet' following centrifugation. Recent work by Marmé and others has shown that irradiation of the tissue with red light before homogenization increases the proportion of phytochrome which is 'pelletable'. Their work suggests that phytochrome in the Pfr form binds readily to membrane fragments. This reversible binding of phytochrome forms the basis of a new technique for concentrating and purifying the pigment.

These results may be an artefact associated with cell homogenization but, on the other hand, they may give a clue to the site of action of Pfr. X, the co-effector for Pfr action, may be a membrane component. It is possible that Pr as well as Pfr is associated with membranes *in vivo* but the former is bound less tenaciously so that it becomes detached during homogenization. The association between phytochrome and a membrane provides a structural basis for the orientation of phytochrome molecules and also for possible co-operative interaction between Pfr molecules. Light induced changes in the phytochrome chromophore leads to changes in the conformation of the phytochrome protein. This in turn could lead to changes in membrane conformation and hence to changes in membrane function.

It should be stressed that permeability change is only one possible consequence of change in membrane function. For instance, there could be changes in the activity of membrane bound enzymes. Release of 'messenger substances' must also be a possibility since the biochemical response may be remote from the site of Pfr action, say in the nucleus of the same cell or in the nucleus or cytoplasm of another cell. In this context the Pfr induced rapid release of gibberellin from etioplasts is of some interest. The variety of responses to Pfr could be associated with a variety of co-effectors $(X_1, X_2, X_3,$ etc.) although it is possible that there is only one co-effector and that the variety of response arises from the states of primary differentiation of the cells concerned.

5·7 Phytochrome and the future

Present knowledge of how light controls plant growth is being put to use in the development of light régimes for the growth of glasshouse crops. Prospects for the future include the possibility of improved growth of field crops and weed control. Although a great deal is now known about the properties and nature of the photoreceptor phytochrome we are far from being able to use this information to explain all the developmental responses of plants to their light environment. The primary mode of action of phytochrome is still unknown although there has been steady progress in filling in the details of the processes that mediate between Pfr action and the final response. The analysis of phytochrome controlled responses coupled with our knowledge of molecular biology will contribute to an increased understanding of the control mechanisms in the development of all higher organisms. A phytochrome controlled response provides an ideal experimental system as the direction of development can be easily changed by a single pulse of light.

6 Some Practical Exercises

6.1 Construction of simple light sources

These light sources should be used in a dark room, preferably maintained at 25°C by means of a fan heater and thermostat.

Red light

Material to be irradiated is placed in a cardboard box with a lid consisting of one layer of No. 14 Ruby and one layer of No. 1 Yellow Cinemoid[1] filter placed beneath two 40 watt fluorescent tubes. Phytochrome photoconversion will be complete after a 5 minutes exposure to this source.

Far-red light

Material to be irradiated is placed in a cardboard box with a lid consisting of one layer of No. 5a Deep Orange and one layer of No. 20 Deep Blue (Primary) Cinemoid filter. The box is placed beneath four 60 watt incandescent lamps and separated from them by a 10 cm depth of water in a glass or clear plastic container. Phytochrome photoconversion will be complete after a 5 minutes exposure to this source.

Green safe light

Fluorescent tube wrapped with at least three layers of No. 39 Primary Green Cinemoid filter.

6.2 Phytochrome control of seed germination

Requirements

Lettuce seeds (variety Grand Rapids[2])
5 cm petri dishes
Filter paper
Cardboard box
Black polythene
Incubator

Method

Sow about 50 seeds of lettuce of the variety Grand Rapids on two layers of filter paper moistened with 2 ml of water in a 5 cm petri dish.

[1] Cinemoid filter obtainable from Rank Strand Electric Ltd., Order Processing Unit, Mitchelstone Estate, Kirkcaldy, Fife, Scotland, U.K.

[2] Grand Rapids lettuce seeds obtained from Thompson and Morgan (Ipswich Seeds Ltd.), Ipswich, Suffolk, U.K.

Immediately place the dishes in a cardboard box wrapped in black polythene and maintain at 25°C in an incubator. The optimum temperature for maximum difference in germination percentage between dark and light treatments varies from one seed batch to another and with the age of the batch. This should be determined by prior experimentation. Use two dishes for each of the following treatments: (i) complete darkness; (ii) 5 minutes red light after 2 hours imbibition and then returned to darkness; (iii) 5 minutes red light followed immediately by 5 minutes far-red light after 2 hours imbibition and then returned to darkness. Manipulations must be carried out in darkness or under a green safelight. After 24 hours count the number of germinated and ungerminated seeds and express the results as percentage germination.

Other experiments

Investigate the dose response relationship (see 4.4) by exposing seeds to various periods of red light (say 15, 30, 60, 120, 240 seconds) after 2 hours' imbibition.

Investigate the time course of escape from far-red reversibility (see 4.8) by maintaining seeds in the dark at 25°C for varying periods of time between the red and far-red irradiations.

6.3 Phytochrome control of seedling growth

Requirements

  Pea seeds
  Vermiculite
  Seed tray
  Test tubes and racks
  Dark room at approximately 25°C

Method

Soak pea seeds for 4 hours in water and then sow in a seed tray containing moist vermiculite. Allow seeds to germinate at 25°C in darkness and after 6 days carefully remove seedlings of uniform size from the vermiculite and place them individually in test tubes containing 10 ml of water. Measure the height of each seedling from the cotyledon to the tip of the epicotyl hook. Use 10 seedlings for each treatment. Treatments are (i) dark control; (ii) 5 minutes red light, then returned to darkness; (iii) 5 minutes red light followed by 5 minutes far-red light, then returned to darkness. Carry out all manipulations under a green safelight. Measure the seedlings again after 24 hours and calculate the mean increase in height for each treatment.

Other experiments

Investigate other aspects of pea seedling photomorphogenesis (see 1.1

and 4.1) using the same procedure. For instance, the effect of red light on the opening of the epicotyl hook and on leaf growth in the apical bud could be studied. The latter effect can be easily quantified by detaching and weighing the apical buds 24 hours after the light treatments.

6.4   Phytochrome control of carbohydrate metabolism

Requirements

    Maize seed
    Vermiculite
    Seed trays
    Beakers
    Boiling tubes
    50 ml volumetric flask
    Pestle and mortar
    Bench centrifuge
    Balance
    Colorimeter
    Dark room at approximately 25°C
    80% ethyl alcohol
    52% perchloric acid
    Iodine solution (2 g iodine and 20 g potassium iodide in one litre)
    Standard starch solution (prepared by dissolving starch in boiling
      water)
    Sand
    Ice

Method

    Soak maize seeds for 4 hours in water and sow in two seed trays containing moist vermiculite. Allow seeds to germinate at 25°C in darkness. On the ninth day expose one tray to 5 minutes red light and then return to darkness. On the tenth day measure the starch content of segments from the middle of the first true leaf. Starch changes take place in both control and irradiated seedlings and the timing of these changes may have to be determined by prior experimentation.

Starch determination: Determine the fresh weight of the leaf material (e.g. fifty 2 cm long median leaf segments). Place the leaf segments in 50 ml of boiling water for 5 minutes, followed by 50 ml of hot 80% ethyl alcohol for 5 minutes. Rinse the segments in distilled water, transfer to a mortar and grind with 5 ml distilled water and a little sand. Transfer the macerate to a boiling tube, rinse the mortar with a further 5 ml of water and add to the boiling tube. Solubilize the starch by adding 15 ml of 52% perchloric acid. *This is a corrosive liquid, so take care to avoid spilling or splashing. Any spilt perchloric acid should be immediately diluted with water, mopped up with a damp*

cloth and the cloth rinsed out in a sink. Dry perchloric acid is potentially explosive.
Place the boiling tube in a bowl of ice and stir occasionally for 15 minutes.
Centrifuge for 5 minutes and decant off the supernatant. Add 7.5 ml of
perchloric acid to the residue and repeat the above procedure. Place the
two perchloric acid extracts in a 50 ml volumetric flask and make up to the
mark with distilled water. Place 10 ml of solution in a colorimeter tube,
add 0.1 ml of iodine solution and determine the amount of blue starch-
iodine complex using a colorimeter with a red filter. Calibrate the
colorimetric assay by using a solution of known starch content.

Other experiments

Investigate the far-red reversibility of red effects on the starch content
of leaves of dark-grown maize seedlings (see 5.1). Also compare the starch
contents of apical, median and basal leaf segments. The red light induced
disappearance of starch takes place more rapidly at the apex of the leaf.
The exercise can be carried out in a simpler although non-quantitative
manner by killing leaves with boiling water, removing chlorophyll with
hot alcohol and then staining for starch in the intact leaf with iodine
solution.

Compare the morphology of seedlings grown in complete darkness
with those treated with red light on the ninth day. Look particularly for
effects on leaf unrolling (see 4.1).

Further Reading

Chapter 1 : Introduction

BORTHWICK, H. A. (1972). History of Phytochrome. In *Phytochrome*, edited by K. Mitrakos and W. Shropshire Jr., p. 4. Academic Press, London and New York.

CLAYTON, R. K. (1970). *Light and Living Matter: A Guide to the study of Photobiology*. Volume 1: *The Physical Part*. McGraw-Hill Book Co., New York.

Chapter 2 : Phytochrome detection and isolation

BRIGGS, W. R. and RICE, H. V. (1972). Phytochrome: Chemical and physical properties and mechanism of action. *Ann. Rev. Plant Physiol.*, 23, 293.

KENDRICK, R. E. and SMITH, H. (1976). The assay and isolation of phytochrome. In *Chemistry and Biochemistry of Plant Pigments*: Volume 2, edited by T. W. Goodwin. 2nd Edition, p. 334. Academic Press, London and New York.

SPRUIT, C. J. P. (1972). Estimation of phytochrome by spectrophotometry *in vivo*: Instrumentation and interpretation. In *Phytochrome*, edited by K. Mitrakos and W. Shropshire Jr., p. 78. Academic Press, London and New York.

Chapter 3 : The properties of phytochrome

FRANKLAND, B. (1972). Biosynthesis and dark transformations of phytochrome. In *Phytochrome*, edited by K. Mitrakos and W. Shropshire Jr., p. 196. Academic Press, London and New York.

HILLMAN, W. S. (1967). The physiology of phytochrome. *Ann. Rev. Plant Physiol.*, 18, 301.

SMITH, H. and KENDRICK, R. E. (1976). The structure and properties of phytochrome. In *Chemistry and Biochemistry of Plant Pigments*: Volume 1, edited by T. W. Goodwin. 2nd Edition, p. 377. Academic Press, London and New York.

Chapter 4 : Phytochrome controlled responses

HILLMAN, W. S. (1967). The physiology of phytochrome. *Ann. Rev. Plant Physiol.*, 18, 301.

FURUYA, M. (1968). Biochemistry and physiology of phytochrome. *Progress in Phytochemistry*, 1, 347.

MOHR, H. (1969). Photomorphogenesis. In *Physiology of Plant Growth and Development*, edited by M. B. Wilkins, p. 489. Academic Press, London and New York.

PRUE, D. (1975). *Photoperiodism in Plants*. McGraw-Hill Book Co., London and New York.

SATTER, R. L. and GALSTON, A. W. (1976). The physiological functions of phytochrome. In *Chemistry and Biochemistry of Plant Pigments*, edited by T. W. Goodwin. 2nd Edition. Academic Press, London and New York.

Chapter 5: Mode of action of phytochrome

MOHR, H. (1972). *Lectures on Photomorphogenesis*. Springer-Verlag, Berlin.

SMITH, H. (1975). *Phytochrome and Photomorphogenesis*. McGraw-Hill Book Co., London and New York.

Proceedings of the 22nd Easter School in Agricultural Science, Nottingham University (1976). *Light and Plant Development* edited by H. Smith. Butterworths, London.